Innovations in Process Engineering

Held at the 2023 AIChE Annual Meeting

Orlando, Florida, USA
5-10 November 2023

ISBN: 978-1-7138-9313-4

Printed from e-media with permission by:

Curran Associates, Inc.
57 Morehouse Lane
Red Hook, NY 12571

Some format issues inherent in the e-media version may also appear in this print version.

Printed with permission by Curran Associates, Inc. (2024)

For permission requests, please contact AIChE
at the address below.

AIChE
120 Wall Street, FL 23
New York, NY 10005-4020

Phone: (800) 242-4363
Fax: (203) 775-5177

www.aiche.org

Additional copies of this publication are available from:

Curran Associates, Inc.
57 Morehouse Lane
Red Hook, NY 12571 USA
Phone: 845-758-0400
Fax: 845-758-2633
Email: curran@proceedings.com
Web: www.proceedings.com

TABLE OF CONTENTS

PROCESS INTENSIFICATION AND MODULAR MANUFACTURING: INTENSIFIED REACTION AND SEPARATION PROCESSES

PROCESS INTENSIFICATION AND MODULAR MANUFACTURING: MODELING AND SIMULATION

BREAKTHROUGHS IN SUSTAINABLE CHEMICAL PRODUCTION AND PROCESS

CONVERSION OF PLASTIC WASTES TO HIGH VALUE PRODUCTS

EFFICIENT PROCESSING OF LIGNIN TO BIOPRODUCTS AND BIOFUELS

IONIC LIQUIDS: NOVEL SEPARATION, CATALYTIC REACTION AND ELECTROCHEMICAL PROCESSES

MATERIALS FOR THERMOCHEMICAL AND ELECTROCHEMICAL ENERGY STORAGE

NEW PROCESSES FOR EFFICIENT CO2 CAPTURE AND UTILIZATION UNDER MILD CONDITIONS

TOPICAL PLENARY: FRONTIERS IN GREEN PROCESS AND PRODUCT ENGINEERING (INVITED TALKS)

PROCESS INTENSIFICATION AND MODULE MANUFACTURING: INTENSIFIED PROCESS FUNDAMENTALS

34a Modular Manufacturing of Active Pharmaceutical Ingredients (APIs): Novel Approaches for Continuous Heavy Metal Removal from API Syntheses Solutions

Daniel Moser, Joscha Boehm, Peter Neugebauer, Heidrun Gruber-Woelfler

Chemical syntheses involving metallic catalysts and reagents are paramount in the pharmaceutical industry. While increasing the efficiency and selectivity of the reaction dramatically, the removal of the often-toxic metal compounds to a level that is harmless for the human body is time, labor, energy, and chemically intensive.

In this contribution, we present the development of modular purification methods for the removal of zinc after a cyclopropanation reaction by means of a Simmons-Smith reaction. The formed cyclopropyl fragment is of significant importance in the pharmaceutical industry as a wide range of active pharmaceutical ingredients (APIs), such as the anti-diabetic drug Saxagliptin, inhere this versatile molecular structure. While the synthesis and the downstream processes are conventionally carried out batch-wise, our project aims to transfer the involved unit operations to continuous mode, including real-time analysis and advanced process control.

The results of the preliminary experiments in batch indicate that a multistage counter-current extraction of the organic product solution with diluted HCl leads to the desired low level of zinc in the product phase. To provide a continuous way of operation at different scales, a multistage setup involving in-house developed 3D-printed continuous stirred tank reactors, continuous membrane separators (Zaiput), and a commercially available extraction column (Sulzer) have been tested.

Moreover, a novel continuous online analysis of the zinc's concentration will be demonstrated. Additionally, the individual pros and cons of both experimental setups are discussed. This includes experimental results, the use of simple-to-measure quality attributes, and the additional impact given by real-time process control. We conclude this discussion with recommendations when which setup might be beneficial.

34c Process Intensification of Electrodialysis through the Investigation and Elimination of Maldistribution

Jack Ledingham, Alasdair Campbell, Jonathan R Howse, Elliot Tyndale-Biscoe, Kyra Sedransk Campbell

Maldistribution in channel-based reactors and separation technologies, such as plate-and-frame heat exchangers, is detrimental to performance through the exacerbation of localised phenomena such as fouling and the reduction of process intensification. Electrodialysis (ED) is an emerging electro separation technology which utilises a stack of ion selective membranes and an electric field to drive salt separation. Industrial implementation and intensification of ED is currently hindered by operational efficiencies being far from ideal. In this work, we explore a system-wide approach to investigate operational inefficiencies resulting from maldistribution between ED channels.

Through computational fluid dynamics (CFD) simulations, it was shown that maldistribution is prevalent in ED. Further, a simplified analytical model demonstrated that the degree of maldistribution will be far more severe upon scale-up. In order to examine the effects of maldistribution on ED performance, transport models were built. These demonstrated that the limiting current density, and thus the maximum salt flux, was reduced by as much as 23% for a standard lab-scale stack. This will be much more significant upon scaleup, and therefore, overcoming maldistribution in ED can have significant benefits for process intensification. Validation of this was conducted through particle image velocimetry (PIV) and desalination experiments.

A glass cell was constructed for PIV experiments with an identical geometry to that used for CFD simulations. A camera and telecentric lens was focussed on the channel centreline, and red and blue LEDs were sequentially pulsed. This light was scattered off particles in water flowing through the cell and collected by a single exposure image. The particle velocities were calculated from the superimposed red and blue images and a flow field was constructed. The experimentally determined flow field had excellent agreement with that found from CFD simulations, validating the existence of maldistribution in ED.

Desalination experiments were conducted on a PC64004 ED stack operating in a steady state mode. The degree of maldistribution was independently altered by changing the flow rate while ensuring that the overall salt convective flux was constant. The LCD was subsequently measured by collecting current-voltage curves. results showed excellent consistency with the transport model, thereby validating the relationship between increased maldistribution and a lower LCD.

In summary, in this work we have demonstrated and validated the presence of maldistribution in ED, as well as the significant detrimental effect it has on the upper limits of process intensification. Further, we have demonstrated the benefits of investigating system-wide inefficiencies in addition to fundamental phenomena. Future research will focus on developing methods to overcome maldistribution in ED.

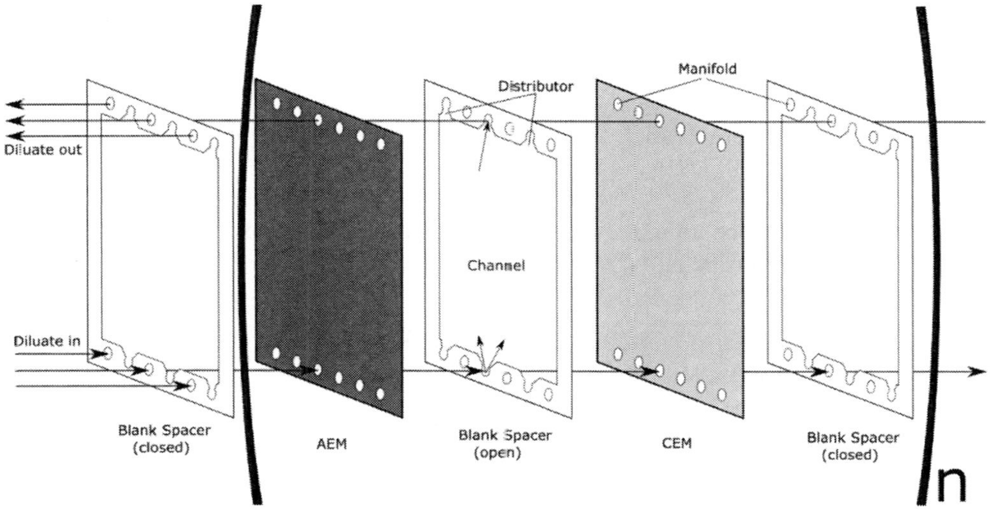
Diluate out
Diluate in
Blank Spacer
(closed)
AEM
Distributor
Channel
Blank Spacer
(open)
Manifold
CEM
Blank Spacer
(closed)
n

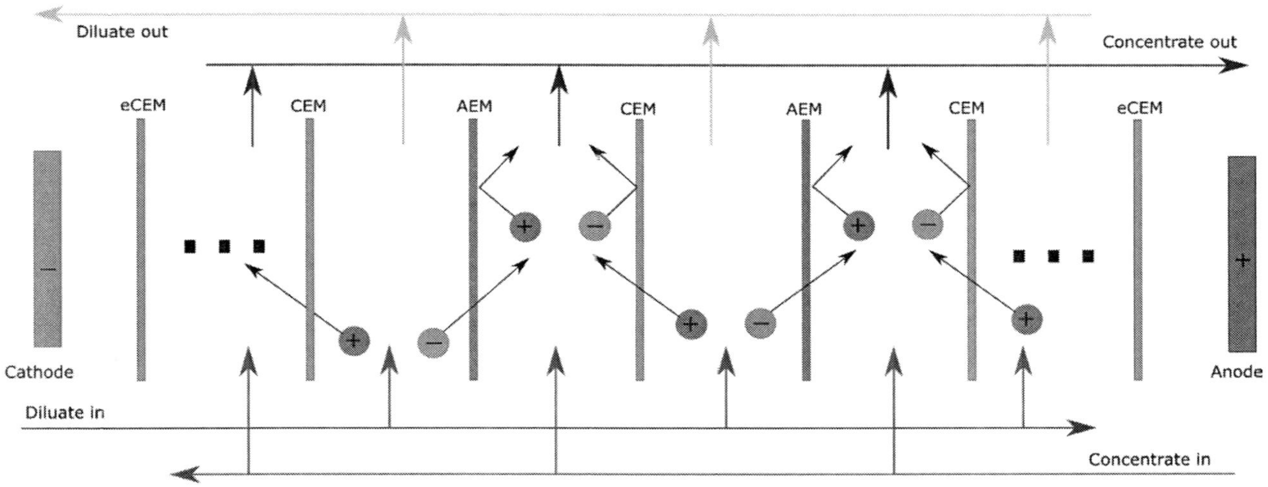
Diluate out
Concentrate out
eCEM
CEM
AEM
CEM
AEM
CEM
eCEM
Cathode
Anode
Diluate in
Concentrate in

34d Intensification of Polymer Production Using Ultrasound: The Polyurethane Case As Example

Ariana Bampouli, Sofie Scheers, Quinten Goris, Joachim Demaerel, Wim De Borggraeve, Georgios D. Stefanidis, Tom Van Gerven

Introduction: A plethora of existing chemical processes has benefited from the implementation of alternative energy sources, such as ultrasound (US). Development of such novel processes is among the most promising and at the same time challenging areas of Process Intensification[1]. Application of US can be found in chemicals and polymers synthesis, in crystallization and in various separation processes, among others[2]. The effects of ultrasound on the production of various chemicals can be physicochemical or mechanical. When a sound wave propagates through a fluid, the violent implosion of numerous gas bubbles is caused, which -locally- triggers extreme temperature and pressure conditions[3]. These bubble collapses termed acoustic *cavitation* result in a series of other phenomena, such as free radical formation, acoustic streaming, high shear stress, mixing in micro-scale, boundary layer disruption and turbulence at very local, small scale[2,4,5]. Focusing on US assisted polymerizations, polymers produced based on the ring opening or free radicals initiation mechanisms have been reported to benefit from the radical species generation caused by US[4]. On the other hand, very few examples exist for polymers produced based on other synthesis mechanisms, such as step growth reactions[6]. Polyurethanes (PUs) –a very important class of polymers– are produced based on the latter mechanism, and are used in a great variety of applications as the final product (elastomers, fibers, foams, etc) or incorporated in a product matrix (e.g. in paints and coatings)[7]. Application of high power US on PUs production reportedly included reaction rate improvement and potential better reactants distribution due to enhanced heating and mixing of the bulk[6]. Nevertheless, the exact mechanism behind these benefits is not yet clarified.

Objective: The goal of the current work is twofold: firstly, obtaining a better understanding of the PU chemistry by thorough investigation of the influence of reaction conditions (temperature, reactants stoichiometric ratio and reactants composition) on the final properties of the PUs produced; secondly, the in-depth investigation of the effects of applying ultrasound for the production of such polymers, compared directly to the conventional thermal processing.

Materials & methods: Various diols were used including ethylene glycol (EG), diethylene glycol (diEG), polyethylene glycol with molecular weight of 200 and 400 g/mol (PEG200 or PEG400). The diisocyanate selected was 4,4-Methylenebis(cyclohexyl isocyanate), referred to as HMDI for simplicity hereafter. HMDI was used as received (mixture of isomers) or for selected cases the trans-trans (termed t,t HMDI herein) form was precipitated, separated, purified and used for the polymerizations. In some reactions bismuth carboxylate was added as the catalyst.

Bulk or solvent-free polymerizations were conducted in small glass vials filled with approximately 10 ml of reactants and two processing modes were applied (Figure 1a): Conventional, or silent, polymerizations using hot oil for the thermal pretreatment and processing of the reactants, or US assisted reactions for which an ultrasonic horn operating at 30 kHz was used in direct contact with the reactants. The reaction temperature was varied between 25 and 140 °C and the reaction progress was monitored based on the isocyanate consumption (decrease of the $N=C=O$ stretching band at 2270 cm^{-1}) using FTIR (Figure 1d). The influence of the reactants stoichiometry was also studied, by varying the ratio of NCO / OH groups. The produced PUs were analyzed using various techniques which included GPC, FTIR, ^{1}H-NMR and DSC. The properties of the final PUs obtained by the two processing modes (Silent and US) were compared and the two processes were evaluated in terms of total energy requirements.

Results: A minimum temperature of 130-150 °C was found to be necessary for the non-catalyzed reactions between EG and HMDI. The effect of the stoichiometric ratio (in molar terms) of the reactants was confirmed to be detrimental for the final PUs molecular weight distribution (MWD) and heat capacity

(Figure 1 b and c). Similarly, the polymeric chain architecture is also determined by the reactants ratio and quality (Figure 1e). When t,t HMDI reacted for example with EG, the *-NH* polyurethane linkage was altered. Both silent and US processing modes resulted in almost identical final products. Additionally, US application seems to decrease the total processing time considerably. Specifically, when sonication of 25 W electrical power (equivalent to 5 W calorimetric power) was applied for the reaction between HMDI and EG (1:2 molar ratio), ten minutes were required for total isocyanate consumption without any additional pretreatment requirements. For the conventional case of the same system, the processing time increased almost by a factor of three including total preheating and reaction time. Both mixing and heating benefits were noticed for the US application in the process. Improved mixing of the two phases formed by the reactants was achieved by complete emulsification of the bulk when sonication was applied. At the same time, depending on the US power applied, very fast heating of the bulk was possible. Experiments in the presence of catalyst revealed that the impact of temperature rise on the PUs production is not straightforward (for both conventional and US processing modes), but it seems that a temperature region exists where degradation of the products is possible. Specifically, at higher temperatures (100 °C) silent products were found to degrade, while degradation appeared at lower temperatures (50 °C) for the US produced polymers.

Conclusions and future steps: In this study, a rather challenging polymerization system is studied, and the reaction is activated using conventional or US treatment. The impact of the reactant quality and stoichiometry is explored, and the final products are characterized using various analytical techniques. The importance of the reactant ratio in the final products obtained is confirmed, and interesting observations are made related to the effect of the reaction temperature selection, when catalyst is present in the system. In addition, it is shown that US can be used to produce polymers of similar properties to the ones produced conventionally. Application of US facilitates both the mixing and the heating of the bulk, and for the reaction between EG and HMDI considerably shorter total processing time is required for the reaction completion. Future steps of this project include additional properties investigation for the PUs produced. Specifically, the final viscoelastic profile of the products will be studied. Further insights in the energy requirements of the two production methods are currently investigated.

References

1. Van Gerven, T. & Stankiewicz, A. Structure, energy, synergy, time-the fundamentals of process intensification. *Ind. Eng. Chem. Res.* 48, 2465–2474 (2009).
2. Sutkar, V. S. & Gogate, P. R. Design aspects of sonochemical reactors: Techniques for understanding cavitational activity distribution and effect of operating parameters. *Chem. Eng. J.* 155, 26–36 (2009).
3. Ashokkumar, M. The characterization of acoustic cavitation bubbles - An overview. *Ultrason. Sonochem.* 18, 864–872 (2011).
4. Suslick, K. S. & Price, G. J. Applications of Ultrasound to Materials Chemistry. *Annu. Rev. Mater. Sci.* 29, 295–326 (1999).
5. Sutkar, V. S., Gogate, P. R. & Csoka, L. Theoretical prediction of cavitational activity distribution in sonochemical reactors. *Chem. Eng. J.* 158, 290–295 (2010).
6. Price, G. J., Lenz, E. J. & Ansell, C. W. G. The effect of high intensity ultrasound on the synthesis of some polyurethanes. *Eur. Polym. J.* 38, 1531–1536 (2002).
7. Akindoyo, J. O. *et al.* Polyurethane types, synthesis and applications – a review. *RSC Adv.* 6, 114453–114482 (2016).

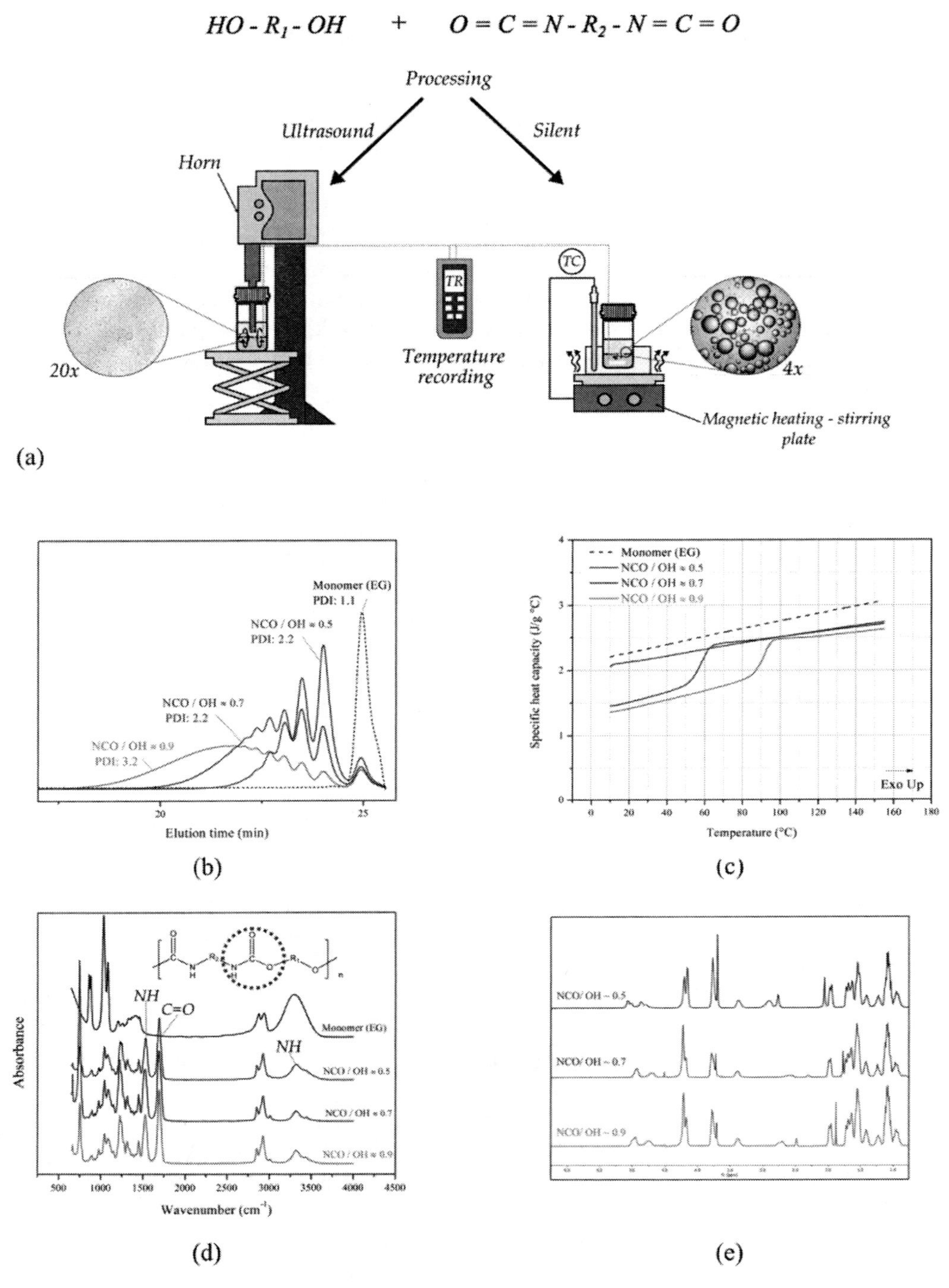

Figure 1: (a) Simplified scheme of the two processing set ups, Ultrasound processing (left) operating at fixed frequency of 30 kHz and power up to 50 W and conventional processing (right) with thermal and magnetic stirring application. In both modes, the temperature of the bulk is recorded online. Products analysis results including: (b) Molecular Weight Distribution (MWD) of various PUs produced with different reactants stoichiometric ratios, the polydispertisty index (PDI) is added in the graphs. (c) Specific heat capacity curves. (d) FTIR analysis. (e) ^{1}H- NMR results (400 MHz, in chloroform–d).

34e Process Design and Intensification of a Continuous Modular Crystallization System

Monika Neal, Montgomery Smith, Zoltan Nagy

Recently there has been a transition from batch to continuous manufacturing to further intensify the standard manufacturing process. Continuous processes in general have shown many promising results such as increased product uniformity, quality, and safety due to minimal human involvement and no batch-to-batch variation, as well as easier scale-up and reduced operating volumes allowing for better control and reduced overall footprint and cost (Su et al., 2019). Using process intensification further improves the robustness and decreases the scale of these processes enabling high degree of automation, modularization, and autonomous operation. The concept of modular manufacturing has gained recognition recently in various academic and industrial communities. Containing continuous manufacturing processes in reconfigurable, automated modular and mobile units can minimize disruptions in the supply chain of various critical products, such as pharmaceuticals and energetic materials, that the current centralized batch manufacturing systems cannot do due to more rigid operation at fixed locations requiring complex logistics for transferring and storing the active ingredients. Modular manufacturing systems are reconfigurable and can be customized to changing process needs.

Current literature has shown few modular manufacturing units, including a Pharmacy on Demand (PoD) portable continuous manufacturing factory containing downstream units for producing battlefield medicines instantly and locally (Capellades et al., 2021). In this work, a modular automated and autonomous continuous crystallization unit is presented consisting of a mixed-suspension mixed-product removal (MSMPR) cascade system. This unit was designed for energetic materials with the motivation of improving personnel safety by eliminating the transport of the energetic compound due to local manufacturing and only requiring raw material transportation. The modular unit is constructed in an intelligent containment unit (ICU) with increased safety including blast-resistant plexiglass, sensors, cameras, and alarms for process monitoring and automated fault detection. It is controlled via custom built LabVIEW software for remote management of all enclosed equipment and allowing the ability to switch between closed-loop (automated design of experiments) and open-loop operation (end-to-end manufacturing). *In-situ* process analytical technology (PAT) tools are integrated in the continuous crystallization system for further control to ensure the desired critical quality attributes (CQAs) are being achieved. Applying this modular design to continuous crystallization achieves robust, efficient, and flexible manufacturing.

1. Su, Q.; Ganesh, S.; Moreno, M.; Bommireddy, Y.; Gonzalez, M.; Reklaitis, G. V.; Nagy, Z. K. A Perspective on Quality-by-Control (QbC) in Pharmaceutical Continuous Manufacturing. *Chem. Eng.* **2019**, *125*, 216–231.
2. Capellades, G.; Neurohr, C.; Briggs, N.; Rapp, K.; Hammersmith, G.; Brancazio, D.; Derksen, B.; Myerson, A. S. On-Demand Continuous Manufacturing of Ciprofloxacin in Portable Plug-and-Play Factories: Implementation and In Situ Control of Downstream Production. Org. Process Res. Dev. 2021, 25 (7), 1534–1546.

101a Modeling and Optimization of Upfront Nitrogen Removal from LNG Baseload Plant Using Lithium Cycle

Fares Almomani, Abdullah Omar

The increase in energy demand as the world population grows, as well as the competition in the Liquefied Natural Gas (LNG) market, force producers to work hard on developing cost-effective production technologies. Upfront Nitrogen Removal (UNR) before the LNG plant's cold section is considered a promising option to save energy that would otherwise be wasted to cool down a large volume of unused nitrogen in the gas stream. In this work, the use of the Lithium Cycle (Li-Cy) as a cost-effective method for UNR is investigated. The Li-Cy is compromised of three stages: lithium chemisorption of nitrogen (, hydrolysis of lithium nitride (and electro-winning (Elec.-w) of the final product to precipitate lithium metal for further reuse. The relevant chemistry, applicability, economic, and future challenges of Li-Cy as an UNR technology from Natural Gas (NG) were explored and discussed. The main challenges which required further investigation to apply Li-Cy to large-scale applications were highlighted for future works. The literature review revealed that Li-Cy can spontaneously remove nitrogen from NG even at low temperatures and produces ammonia as a valuable hydrogen storage material. The used lithium can be regenerated via and Elec.-w and reused again many times. The cost of the Li-Cy can be compensated by energy savings, the increase in production rate, and by selling the generated ammonia. Calculations showed that selling the produced ammonia from LNG plants with capacity in the range of 1 to 5 MTPA would not only offset the costs of Li-Cy but would generate a net profit of $21MM to $103MM, respectively.

101b Thermodynamic Modeling of CO_2 Absorption in K_2CO_3-H_2o Solution with Association Enrtl Model

Cheng-Ju Hsieh, Viraj Patil, Chau-Chyun Chen

An environmentally benign absorbent, K_2CO_3-H_2O solution is a common chemical absorbent for CO_2 capture due to its low cost. Recently, the industry is gaining interest in developing high-temperature and high-concentration K_2CO_3-based CO_2 absorption processes to reduce energy requirement associated with CO_2 stripping and compression processes. Previous thermodynamic modeling research has not investigated enthalpy of CO_2 absorption and vapor-liquid equilibrium, especially water vapor pressure, at high temperatures. This study expands the prior thermodynamic modeling of the K_2CO_3-H_2O-CO_2 ternary system and its associated binary systems with a temperature range of 273.15 – 473.15 K and the K_2CO_3 concentration up to saturation with the association electrolyte nonrandom two-liquid (association eNRTL) model (Y.-J. Lin, C.-J. Hsieh, and C.-C. Chen, *AIChE Journal*, 68 (2022) e17422). The association eNRTL model is well suited for thermodynamic modeling of electrolyte solutions with high charge density ions with strong association tendency including the carbonate ions. Accounting for both self- and cross-association of water and carbonate ions, the association model provides an improved representation of the K_2CO_3-H_2O-CO_2 ternary system compared to the original eNRTL model.

101c Optimal Operation of Intensified Fluidized Bed Membrane Reactor for Oxidative Coupling of Methane

Moustafa Ali, Dustin Kenefake, Yuhe Tian, Efstratios Pistikopoulos

The catalytic oxidative coupling of methane (OCM) process has received intense interest in the literature due to the potential to directly convert natural gas or methane to value-added chemicals at a reduced cost, energy consumption and carbon emissions when compared to conventional processes. However, major challenges such as low yield, catalyst deactivation, and reactor scale-up still challenge the commercialization of this process (Galadima & Muraza, 2016; Onoja et al., 2019). A promising solution to address these challenges is to develop innovative OCM reactor designs leveraging the recent advances in modular process intensification, e.g. membrane reactors (Cruellas et.al 2020). While the investigation of OCM reaction systems has mostly been focused on catalyst development and steady-state conceptual designs, ignoring the process dynamics limits the applicability of evaluating the implementation for commercial use (Barteau, 2022). The aim of this work is to design a dynamic optimal OCM process at commercial scale leveraging the concept of modular process intensification.

In this work, we present the new optimal intensified fluidized bed membrane reactor (FBMR) design system for the simulation of oxidative coupling of methane. The conventional fluidized bed reactor was simulated and validated with a C_2+ yield, selectivity, and methane conversion of 15%, 54% and 26% respectively. A membrane was integrated with the conventional reactor by coupling a perovskite oxygen selective membrane along the axial direction to improve on the three key performance indicators mentioned above. The use of membrane for oxygen feed distribution has been reported to result in better C_2+ yield and selectivity by selectively enhancing the desired reactions (Spallina et al., 2015). The FBMR reactor's design shows a noticeable improvement with a C_2+ yield of 35%, C_2+ selectivity of 67% and improved conversion of methane up to 50% with a reduced CO_2 production by almost 81% (Ali et.al, 2023). Although the new proposed intensified FBMR design solves some of the limitation for the OCM reaction to take place, it opened few research questions to uncover the role of membrane in achieving high C_2+ selectivity%, C_2+ yield% and CH_4 conversion%. The effect of these parameters (e.g. oxygen vacancy dilution coefficient (D_V), species diffusion coefficient (D_i) and perovskite membrane mass transfer coefficients (k_f & k_a)) are investigated to enhance the economics of the OCM reactions and the performance of FBMR. Finally, given the large-scale highly nonlinear nature of the high-fidelity model, a linearized surrogate model is built to balance the computational complexity for model-based control execution and the model accuracy for predictive optimization. The optimal design and operational policy for the FBMR is developed using the PAROC (PARametric Optimization and Control) framework (Pistikopoulos et al., 2015; Pappas et al., 2021). Then, the multi-parametric model predictive control (mp-MPC) problem is solved using the POP toolbox (Pistikopoulos et al., 2020), allowing to analytically derive the optimal explicit operation policy as an affine function of state variables, disturbances, design variables, etc. resulting in enhanced C_2+ yield, selectivity, and CH_4 conversion with guaranteed operational feasibility under disturbances

Keywords: OCM, Process Intensification, Fluidized Bed Membrane Reactors

References

Ali, M.; Tian, Y; Kenefake,D ; Pistikopoulos, E. N. Process Design and Intensification of Circulating Catalytic Fluidized Bed Membrane Reactor for Oxidative Coupling of Methane. ESCAPE-33

Barteau, M. A. (2022). Is it time to stop searching for better catalysts for oxidative coupling of methane? Journal of Catalysis, 408, 173–178.

Cruellas, A., Melchiori, T., Gallucci, F., & van Sint Annaland, M. (2020). Oxidative Coupling of Methane: A Comparison of Different Reactor Configurations. Energy Technology, 8(8), 1–15.

Galadima, A., & Muraza, O. (2016). Revisiting the oxidative coupling of methane to ethylene in the golden period of shale gas: A review. JIEC, 37, 1–13.

Onoja, O. P., Wang, X., & Kechagiopoulos, P. N. (2019). Influencing selectivity in the oxidative coupling of methane by modulating oxygen permeation in a variable thickness membrane reactor. Chem.Eng.Process, 135(2018), 156–167.

Pistikopoulos, E. N., Diangelakis, N. A., Oberdieck, R. (2020). Multi-Parametric Optimization and Control. Vol.1.

Pistikopoulos, E. N., Diangelakis, N. A., Oberdieck, R., Papathanasiou, M. M., Nascu, I., & Sun, M. (2015). PAROC -- An integrated framework and software platform for the optimisation and advanced model-based control of process systems. Chemical Engineering Science, 136, 115–138.

Pappas, I., Kenefake, D., Burnak, B., Avraamidou, S., Ganesh, H. S., Katz, J., Diangelakis, N. A., & Pistikopoulos, E. N. (2021). Multiparametric Programming in Process Systems Engineering: Recent Developments and Path Forward. Frontiers in Chemical Engineering, 2(January), 1–15.

Spallina, V., Melchiori, T., Gallucci, F., & Van Sint Annaland, M. (2015). Auto-thermal reforming using mixed ion-electronic conducting ceramic membranes for a small-scale H2 production plant. *Molecules*, *20*(3), 4998–5023.

101d Techno-Economic Optimization of Heat-Integrated Multi-Functional System of Catalytic Reactors: The Case of Water-Gas Shift and COS Hydrolysis Reactions

Pranav Kherdekar, Shantanu Roy, Divesh Bhatia

1. Introduction

The design of the solid carbonaceous fuel-to-methanol production process needs to be carried out so as to reduce the production cost of methanol. However, the syngas derived from solid carbonaceous fuels contains multiple types of contaminants. Further, the relative concentrations of H_2, CO, and CO_2 present in the syngas are not ideal for obtaining maximum methanol yield from the syngas. Thus, a network of multiple processes is required for the clean-up and conditioning of the syngas. Due to the presence of multitude of sub-processes, syngas clean-up and conditioning unit results in an increase in the cost of production of methanol. The overall economy of the methanol production process can be improved by applications of principles of process intensification, as demonstrated by Stankiewicz [1], to individual parts of the process so as to reduce the capital cost and operating expenses. In this study, we demonstrate the intensification of catalytic systems for water-gas shift and COS hydrolysis reactions by treating it as a techno-economic optimization problem.

In a commercial plant, the water-gas shift reaction is carried out using H_2S-containing syngas with Co-Mo-based sour shift catalysts in a fixed-bed reactor. The COS hydrolysis reaction, on the other hand, is carried out over metal oxides such as γ-Al_2O_3 in a separate fixed-bed reactor [2]. However, it is noteworthy that the γ-Al_2O_3 support used in the Co-Mo/γ-Al_2O_3 has the capability to catalyse the COS hydrolysis reaction as well [3]. Further, both water-gas shift and COS hydrolysis are exothermic reactions requiring the same co-reactant, i.e., steam (H_2O). Hence, the possibility of carrying out both the reactions in a single multifunctional reactor rather than a system of multiple catalytic reactors needs to be assessed. Further, the overall catalytic process requires multiple heating and cooling arrangements leading to high capital and operating costs. The requirement of steam for the two reactions further increases the operating costs. To reduce the adverse impact of the catalytic processes on the overall economy of the methanol production process, optimization of the catalytic system needs to be carried out with an objective of minimizing the operating and capital costs.

2. Methodology

A one-dimensional heterogeneous model is developed for an adiabatic fixed-bed reactor. The model is adapted to simulate a single fixed-bed reactor loaded with Co-Mo/Al_2O_3 catalyst in which both the water-gas shift and COS hydrolysis reactions take place. The kinetics of the two reactions are adopted from the literature [2,4]. For optimization, the 1-dimensional reactor model is coupled with the elitist population-based non-dominated sorting genetic algorithm (NSGA-II). CO conversion of 40% is found to be required to attain the stoichiometric number of 2, which is considered optimal for methanol production. The minimization of deviation of the CO conversion (X_{CO}) from the required value ($X_{CO,req}$) is considered as the first objective of optimization. The maximization of the conversion of COS (X_{COS}) is considered as the second objective of the optimization problem. To address the problem of techno-economic optimization, minimization of the product of the capital costs and operating costs (CAPEX*OPEX) is considered as the third objective of the optimization. The capital costs are calculated as the sum of the cost of the fixed-bed reactor vessel including the catalyst and the cost of the heat exchangers required for a minimum energy target configuration of the heat-exchanger network [5]. Further, the operating costs considered include the cost of process and utility steam, cooling water, degradation of the guard bed placed downstream, and the cost incurred due to pressure drop. The cost of the utilities is calculated for the area-targeted heat exchanger network. The three-objective optimization problem is solved and the resulting Pareto fronts are analyzed. Based on the results, a heat-integrated reactor system capable of carrying out both water-gas shift and COS hydrolysis reactions with low capital and operating costs is proposed.

3. Results and discussion

Fig. 1 (a) shows the Pareto-optimal front obtained after solving the three-objective optimization problem. It is observed that at a constant value of deviation from the required conversion of CO (x-axis), the value of CAPEX*OPEX (contour) increases with an increase in the conversion of COS (y-axis). It is further observed that the maximum COS conversion possible in the reactor decreases as CO conversion becomes close to the required value. This can be explained based on the variation of the conversions of CO and COS with the feed temperature (Fig. 1 (b)). It can be noted that high temperatures correspond to lower deviation from the required CO conversion but also result in low COS conversions. The water-gas shift reaction with an appreciable conversion is favored at relatively higher temperatures (>280°C). At these temperatures however, the equilibrium limitations on the COS hydrolysis reaction set in and the reversible reaction dominates. This is because water-gas shift and COS hydrolysis are both reversible exothermic reactions and the activation energy for COS hydrolysis on the Co-Mo/γ-Al_2O_3 is comparatively lower than that for the water-gas shift reaction. Thus, increasing the temperature increases the conversion of CO, as the reaction operates in the kinetic regime, while increasing the temperature reduces the conversion of COS, as the reaction operates in equilibrium-limited regime.

Fig. 1 (c) shows the variation of the COS and CO conversions with the steam to CO ratio. It is observed that at a constant deviation from the conversion of CO (x-axis), the conversion of COS increases with an increase in the steam to CO ratio. This trend can also be explained on the basis of thermodynamic limitations on the COS conversion at the reactor operating temperatures. Here, higher steam to CO ratios increase the partial pressure of reactant (steam) thereby increasing the rate of forward COS hydrolysis reaction. However, comparison of Figs. 1 (a) and (c) shows that the chromosomes associated with higher steam to CO ratio are also associated with high values of CAPEX*OPEX. This is because process steam contributes significantly (60-85%) to the operating costs of the multi-functional catalytic system. Fig. 1 (d) shows the variation of the contribution of the cost of steam to the total operating costs of the reactor with the conversion of COS. Higher conversion of COS is found to be linked with higher steam costs. Cost incurred due to pressure drop, ZnO (guard bed) degradation, and cooling water are found to have a relatively lower contribution (0-25% each) to the operating costs. Fig. 1 (e) shows the variation of CAPEX and the contribution of heat exchangers to the CAPEX with the conversion of COS. It is observed that high conversion of COS (>95%) is linked to high CAPEX. On the other hand, this effect is not observed for relatively lower values of conversions of COS (85-90%). The heat exchangers (HE) contribute the highest to the CAPEX (40-70%) while reactor fabrication and catalyst costs have relatively lower individual contribution to the CAPEX (5-35%). In contrast to the heat exchanger costs, the combined reactor and catalyst costs are found to decrease with an increase in the COS conversion.

Fig. 1 (f) shows the final reactor design chosen based on the CAPEX and OPEX trends on the Pareto-optimal front and corresponds to 0.1% deviation from the required CO conversion while providing a COS conversion of 86.5%. For this design, the pinch method suggests no requirement of the hot utility since the heat surplus generated by the water-gas shift reaction is sufficient to heat the reactants. Thus, the techno-economic optimization performed in the present study provides a multi-functional reactor design for water-gas shift and COS hydrolysis reactions with lower number of equipment than the conventional process. It further filters the designs that lead to lower capital and operating costs while considering the hot and cold composite curves. Thus, by combining a detailed reactor model, economic analyses and principles of heat integration, a methodology is presented for obtaining the most optimal design for a system of exothermic reactors with multiple heating and cooling arrangements. The full paper will provide an in-depth discussion on other optimization runs that were carried out with the conventional reactor system, the comparison of the optimum designs of the conventional system with the multi-functional reactor, as well as the impact of other decision variables on the reactor performance and costs.

4. Conclusions

A single multi-functional fixed-bed reactor capable of catalyzing both water-gas shift and COS hydrolysis reactions is optimized while considering the capital and operating costs. The optimized multi-functional reactor operates with lesser number of equipment than the conventional system and with lower capital and operating costs. Overall, a methodology of techno-economic optimization of a heat-integrated system of exothermic catalytic reactor systems is presented.

References

[1] A.I. Stankiewicz, J.A. Moulijn, Chem. Eng. Prog. (2000) 22–34.

[2] S. Tong, I.G. Dalla Lana, K.T. Chuang, Can. J. Chem. Eng. 73 (1995) 220–227.

[3] Z.M. George, 271 (1974) 261–271.

[4] S.S. Hla, G.J. Duffy, L.D. Morpeth, A. Cousins, D.G. Roberts, J.H. Edwards, Int. J. Hydrogen Energy. 36 (2011) 6638–6645.

[5] B. Linnhoff, E. Hindmarsh, Chem. Eng. Sci. 38 (1983) 745–763.

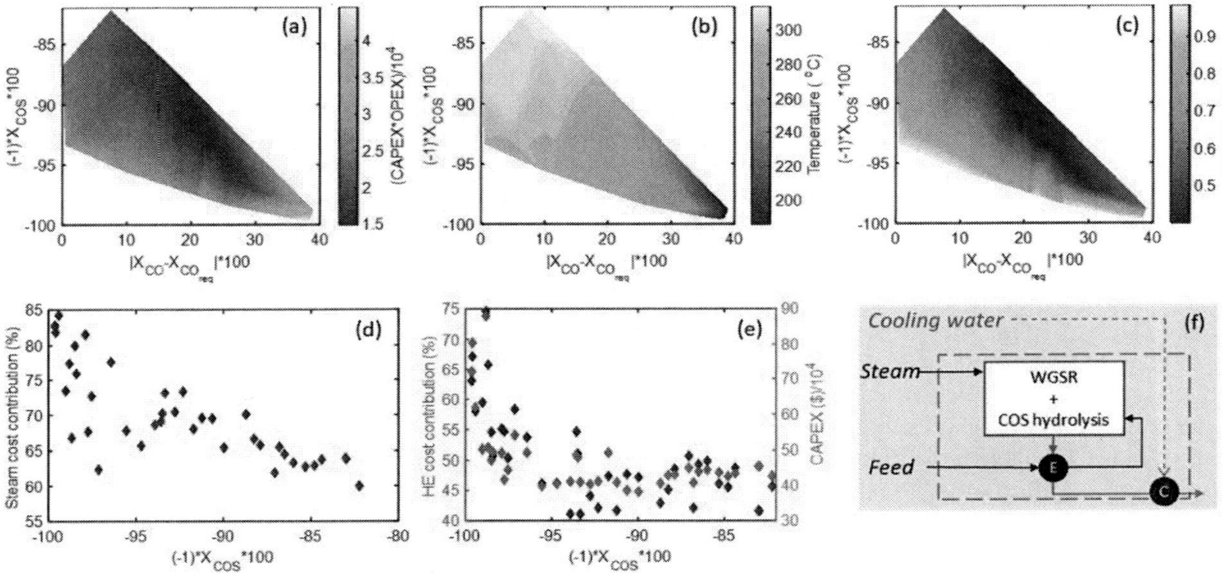

Fig. 1. Results of the three-objective optimization of multi-functional reactor (a) Pareto-optimal front, (b) variation of CO and COS conversion with feed temperature, (c) variation of CO and COS conversion with steam to CO ratio (d) contribution of steam to operating costs (e) contribution of heat exchangers to capital costs (f) Final optimized configuration

101e Machine Learning Aided Process Design and Intensification Using Sparse Experimental Data: An Ammonia Production Example

Md Al Masud, Yuhe Tian

With the burgeoning growth of computing power and algorithmic advances, machine learning (ML) has become a powerful tool to identify multivariate relationships and find statistical patterns behind large amounts of data. ML has also been extensively leveraged to optimize chemical process design and operations by generating reduced order surrogate models from well-established mechanistic models [1-2] and/or enhancing the solution efficiency of sophisticated optimization algorithms [3]. A research question lies in how ML techniques can assist the development of new process designs at the early experiment stage, during which prior process knowledge and experimental data are both limited [4]. The lack of training data presents formidable challenges to generalize the analytics to new test data inputs due to overfitting.

In this work, we present a ML-aided process design strategy with application to a novel microwave-assisted ammonia production process [5]. Current commercial ammonia production routes via thermal Haber-Bosch process feature very high energy intensity at high temperature and pressure. Unconventional microwave-assisted catalytic ammonia synthesis offers the unique advantage to achieve high reaction activity under moderate operating conditions via selective heating [6-7]. A set of experiment data is collected totaling 46 data points to comprise the training and test data sets. Four input features are included: pressure, temperature, hydrogen to nitrogen ratio, and feed flow rate. The goal of process design is to maximize the resulting ammonia concentration over a given amount of CsRu/CeO2. We investigate and compare three classes of methodologies: (i) Statistical analysis using Response Surface Methodology (RSM), in which a polynomial input-output function is correlated for use in optimization [8], (ii) Artificial Neural Network (ANN), in which a regression model is developed using the limited training data set [9], (iii) Deep neural network, in which synthetic minority oversampling technique is further utilized to generate new data samples following the original data distribution to improve prediction accuracy [10]. Optimal process design conditions suggested by the aforementioned computational methods are compared with the designs optimized via experimental analysis. Extensions of the methodologies will also be discussed to unravel the characteristics of microwave-assisted catalytic processes.

References

[1] Cozad, A., Sahinidis, N. V., & Miller, D. C. (2014). Learning surrogate models for simulation-based optimization. AIChE Journal, 60(6), 2211-2227.

[2] Kim, S. H., & Boukouvala, F. (2020). Machine learning-based surrogate modeling for data-driven optimization: a comparison of subset selection for regression techniques. Optimization Letters, 14(4), 989-1010.

[3] Wilson, Z. T., & Sahinidis, N. V. (2017). The ALAMO approach to machine learning. Computers & Chemical Engineering, 106, 785-795.

[4] Cao, B., Adutwum, L. A., Oliynyk, A. O., Luber, E. J., Olsen, B. C., Mar, A., & Buriak, J. M. (2018). How to optimize materials and devices via design of experiments and machine learning: Demonstration using organic photovoltaics. ACS nano, 12(8), 7434-7444.

[5] Wang, Y., Khan, T. S., Wildfire, C., Shekhawat, D., & Hu, J. (2021). Microwave-enhanced catalytic ammonia synthesis under moderate pressure and temperature. Catalysis Communications, 159, 106344.

[6] Muley, P. D., Wang, Y., Hu, J., & Shekhawat, D. (2021). Microwave-assisted heterogeneous catalysis.

[7] Chen, W., Malhotra, A., Yu, K., Zheng, W., Plaza-Gonzalez, P. J., Catala-Civera, J. M., & Vlachos, D. G. (2021). Intensified microwave-assisted heterogeneous catalytic reactors for sustainable chemical manufacturing. Chemical Engineering Journal, 420, 130476.

[8] Evans, M. D., Ring, J., Schoen, A., Bell, A., Edwards, P., Berthelot, D., & Baldino, C. M. (2003). The accelerated development of an optimized synthesis of 1, 2, 4-oxadiazoles: application of microwave irradiation and statistical design of experiments. Tetrahedron letters, 44(52), 9337-9341.

[9] Golkarnarenji, G., Naebe, M., Badii, K., Milani, A. S., Jazar, R. N., & Khayyam, H. (2019). A machine learning case study with limited data for prediction of carbon fiber mechanical properties. Computers in Industry, 105, 123-132.

[10] Liu, X., Xu, Y., Li, J., Ong, X., Ibrahim, S. A., Buonassisi, T., & Wang, X. (2021). A robust low data solution: dimension prediction of semiconductor nanorods. Computers & Chemical Engineering, 150, 107315.

101f Effect of Ethylene Oxide Reactor Intensification on the Downstream Processes

Chinmoy Basak Mukta, Selen Cremaschi, Mario Eden

Reactor intensification in the chemical process can significantly impact downstream separation, particularly for processes like ethylene oxide (EO) production, where exothermic reactors handle a large amount of inert and unwanted byproducts (i.e., carbon dioxide). The byproducts must be removed before the unreacted reactant can be recycled back into the reactor [1]. Intensifying such a reactor, on the one hand, may provide the opportunity for an increase in reactor efficiency and throughput, but the realization of the efficiency depends on the existing downstream separation process, which may or may not be able to handle the separation.

Reactor intensification with Microfibrous Entraped Catalysts (MFECs) has gained considerable interest in academia and the industry alike. MFECs are a type of catalyst enhancer composed of microfibers metals that entraps catalytic particles and typically have diameters in the range of a few micrometers to a few hundred micrometers [2]. One of the key advantages of MFECs is their high surface area-to-volume ratio, which allows for a greater concentration of catalytic particles per unit volume compared to traditional catalysts. This higher concentration can lead to increased catalytic activity and selectivity, as well as more efficient use of the catalytic material. Another advantage of MFECs is their ability to provide good mixing and convective heat transfer characteristics due to the small size, high surface area, and higher thermal conductivity of the microfibers. This can improve EO reaction kinetics and reduce the potential for mass and heat transfer limitations of traditional catalysts [3-5].

Re-routing or reconfiguration of the chemical process flowsheet by utilizing unused separation columns can be an effective way to enhance separation capacity. This approach involves identifying and utilizing underutilized or idle separation columns in the process flow and reconfiguring the flow to use these columns for other separations instead of investing in new equipment. Previous research works have explored the potential benefits of reconfiguration in terms of cost savings and energy efficiency. As a result of reactor intensification using MFECs, inert materials were reduced, leading to a decrease in solvent usage in the EO separation column. This decrease in solvent use subsequently eliminated the need for one of the two EO purification columns, which can be repurposed to handle excess capacity, resulting in an overall increase in the efficiency of the separation process and process cost reduction.

Derivative-free optimization (DFO) is a framework for solving optimization problems where the objective function and/or its derivative cannot be expressed mathematically. Instead, DFO uses algorithm-specific search strategies to improve objective function values for sets of decision variables until the best set is found [6]. DFO treats the model used to estimate objective function values as a black-box model. In this work, we have applied the DFO framework to find the optimized flowsheet to minimize the overall cost of EO production.

Overall, in this work, reactor intensification with MFEC has been shown to improve the efficiency of the EO separation process. In addition, re-routing or reconfiguring the process flowsheet can reduce operating costs by up to 22.3% for an existing brownfield process and up to 37.5% for a greenfield design, compared to the conventional process. These cost reductions can be achieved by identifying and utilizing underutilized or idle separation columns.

References

[1] Mukta, C.B., Cremaschi, S., Eden, M. R., Tatarchuk, B. J. (2022). Techno-Economic Study of Intensified Ethylene Oxide Production Using High Thermal Conductivity Microfibrous Entrapped Catalyst. Computer Aided Chemical Engineering (Vol. 51, pp.1 Elsevier.

[2] Cheng, X., Yang, H., & Tatarchuk, B. J. (2016). Microfibrous entrapped hybrid iron-based catalysts for Fischer–Tropsch synthesis. Catalysis Today, 273, 62-71.

Directive, E. U. (2005). 36/EC of the European Parliament and of the Council of 7 September 2005 on the recognition of professional qualifications. Off J Eur Union, 22-142.

[3] Sheng, M., Gonzalez, C. F., Yantz Jr, W. R., Cahela, D. R., Yang, H., Harris, D. R., & Tatarchuk, B. J. (2013). Micro scale heat transfer comparison between packed beds and microfibrous entrapped catalysts. Engineering Applications of Computational Fluid Mechanics, 7(4), 471-485.

[4] Kalluri, R. R., Cahela, D. R., & Tatarchuk, B. J. (2009). Comparative heterogeneous contacting efficiency in fixed bed reactors: Opportunities for new microstructured systems. Applied Catalysis B: Environmental, 90(3-4), 507-515.

[5] Cheng, P., & Tatarchuk, B. J. (2019). Kinetic study of SO2 adsorption on microfibrous entrapped sorbents for solid oxide fuel cell cathode protection. Chemical Engineering Science, 201, 157-166.

[6] Boukouvala, F., Misener, R., & Floudas, C. A. (2016). Global optimization advances in mixed-integer nonlinear programming, MINLP, and constrained derivative-free optimization, CDFO. European Journal of Operational Research, 252(3), 701-727..

139a Identifying Molecular Descriptors for CO_2 Chemisorption in Eutectic Mixtures Using High-Throughput Screening

Stephen Vicchio, Rachel Getman

Designing new materials for the direct air capture (DAC) of atmospheric CO_2 is essential to limit global temperature rise below 1.5 °C by 2050, with these new negative emission technologies (NETs) required to remove an estimated 10 gigatonnes of CO_2 annually. These technologies necessitate discovering new materials that can reversibly bind CO_2 under diverse environmental conditions with minimal energy requirements. Eutectic mixtures are one such class of materials that exhibit excellent properties for DAC. Eutectic mixtures consist of hydrogen bond acceptors (e.g., halide salts) and hydrogen bond donors (e.g., alcohols, amines, or amides) that together form a complex, dynamic liquid structure. Prior research highlights how CO_2 uptake performance is tunable by changing the components of the eutectic mixtures, allowing for CO_2 uptake optimization. However, at present, the design rules governing uptake remain unknown for these materials, which inhibits rational materials development. Therefore, we utilize high-throughput screening to understand the structure-function relationship behind CO_2 chemisorption in different eutectic mixtures. Specifically, we use quantum chemical calculations to screen different combinations of hydrogen bond acceptors and hydrogen bond donors. We compute the CO_2 binding and hydrogen dissociation energies for each species in our libraries, and then construct thermodynamic pathways for CO_2 binding and proton transfer reactions (as both reactions are fundamental to DAC in eutectic mixtures). Ultimately, this work identifies key descriptors that are important for CO_2 chemisorption. Our findings have significant implications for experimental synthesis, characterization, and performance evaluation of new eutectic mixtures for DAC, leading to the rational design of new eutectic mixtures with optimal CO_2 absorption. Ultimately, filling essential gaps in eutectic mixtures performance for DAC makes them a more viable NET.

139b Carbon Dioxide Capture and Utilization Technology for Recycling Heavy Metals from Industrial Metal Wastes

Sayan Banerjee, Asad Abbas, Vikram Singh, K.D.P. Nigam

Uncontrolled release of anthropogenic CO_2 emissions from power plants and other industries (e.g., cement, steel production, natural gas production, etc.) is causing an alarming situation, like global climate change. Due to global warming, an increase in sea level of 38 cm and 1.9 °C in global mean temperature has been predicted by the year 2100. Scientists are trying to develop efficient and cost-effective carbon capture and sequestration technologies to combat these climate changes. Although carbon capture and storage (CCS) is the conventional method presently used, the processes of transportation of CO_2 to the storage site, injection of CO_2 in the rock pores, and storage depends on complicated machinery and suitable underground reservoirs. In addition, the process poses the risk of accidental release of CO_2 due to seepage. Carbon capture and utilization (CCU) is currently the preferred option as it can mitigate global climate change by converting the captured CO_2 into value-added chemicals such as chemicals and fuels at the same time. The present work aims to develop disruptive technology for capturing and utilizing carbon dioxide to recycle heavy metals (e.g., copper, zinc, nickel, cobalt, etc.). The added advantage of our current technology lies in reducing the carbon footprint from mining and utilizing other secondary sources of metals like e-wastes and other industrial wastes. E-waste (e.g., PCBs, copper wires, etc.) is also referred to as urban mines as it is a rich source of many valuable metals. Unplanned discard of these hazardous wastes has a negative impact on the environment as well as human health. The objective of the present work is to recycle e-wastes and other hazardous industrial wastes using carbon dioxide from flue gas and extract the valuable metals in metal carbonate, a stable form of metal salts. About 97% zinc extraction has been achieved in the present work from an industrial zinc waste, utilizing captured CO_2 using ammonia solution in a scrubber. Zinc has been extracted as zinc carbonate, having numerous industrial applications, like catalyst synthesis, rubber production, personal cleanliness products, petroleum industry as a sulfur absorber, and feed additives for animals.

139c Electrochemical Regeneration of High Purity CO_2 from Carbonates for Efficient Self-Looped Carbon Capture

Zhiwei Fang, Haotian Wang

Capturing and concentrating low-content CO_2 in air is a crucial step for carbon neutralization and utilization. The electrochemical approach, using renewable energy as input, is a green and sustainable way to concentrate atmospheric CO_2, but is still challenged by low stability and low purity of as-captured CO_2. Herein, inspired by the charge-discharge process of chargeable batteries, we developed a module system to capture the CO_2 directly from the air and regenerate pure CO_2 through an electrochemical process. Inspired by the charge-discharge process of chargeable batteries, we proposed a close-looped process including direct air CO_2 capture, CO_2 release and the simultaneous regeneration of alkaline solutions for subsequent CO_2 absorption/desorption cycle. The three-chamber electrochemical reactor with a permeable ion-conducting solid electrolyte layer is applied, using renewable energy as input, would be a green and sustainable method to enrich atmospheric CO_2. When direct supplying carbon-containing bicarbonate/carbonate solution, a high efficiency of over 90% for hydroixde, as CO2 absorber, can be achieved. By continued cycling of the carbon-containing solution, 99% purity CO_2 gas was continuously generated with a CO_2 collection efficiency of more than 80% for over 100 h. Our design is compatible with today's carbon neutrality aim those functions optimally with high efficiency and large compacity. The electrochemical approach, using renewable energy as input, would be a green and sustainable method to enrich atmospheric CO_2, and other materials.

139d Effect of Hydrodynamics on Microalgal Growth in Externally-Illuminated Tubular Bubble-Column Photobioreactors

Gourab Ghosh, Arnab Atta

This study focuses on how the kinetics of microalgal growth in a bubble-column photobioreactor are affected by radial recirculation. We employ an experimental setup to show how the growth dynamics change over time for two distinct flow patterns in a bubble column photobioreactor: one that only allows for axial air flow and another that allows for adequate radial algal movement. Our experiments have been performed using extremophillic microalga *Chlorella sorokiniana* for 6-8 days in a 5 L externally lit bubble-column photobioreactor, with various CO_2 concentrations, ranging from atmospheric condition (0.04% v/v) through 1%, 2%, 3%, 4% to 5% concentrations. External illumination was kept at 10500 lux throughout the course of the experiment.

Our investigation demonstrates that the well-circulated radial flow pattern inside a bubble column cylindrical photobioreactor encourages higher microalgal yield compared to its counterpart at lower CO_2 concentrations (atmospheric to 2% condition). Both nitrogen-limited and non-nitrogen-limited instances were used to verify our assertions, and we discovered that both cases showed similar findings. The microalgal biomass content was found to be 4.62 and 5.40 g/L after 186 hours in radially-mixed nitrogen limited and non-nitrogen limited instances, respectively, as opposed to 4.13 and 4.87 g/L in its counterparts. However, at higher CO_2 concentrations, the algal media's pH, in the radially well-mixed configuration, declines rapidly to below 7.5, which reduces the algal productivity. The failure of the radial bubbles at higher algal density to drive them across the reactor radii to the periphery of the externally-lit reactor, as well as the lower pH, may be responsible for the lower average algal concentration in the radially well-mixed scenario at higher CO_2 concentrations.

As a result, we advocate greater radial mixing at lower CO_2 levels (<3% CO_2). Conversely, axial mixing is preferred for greater algal output at higher CO_2 input concentrations.

Keywords: Bioprocessing, Bioenergy, Microalgae, Green Energy

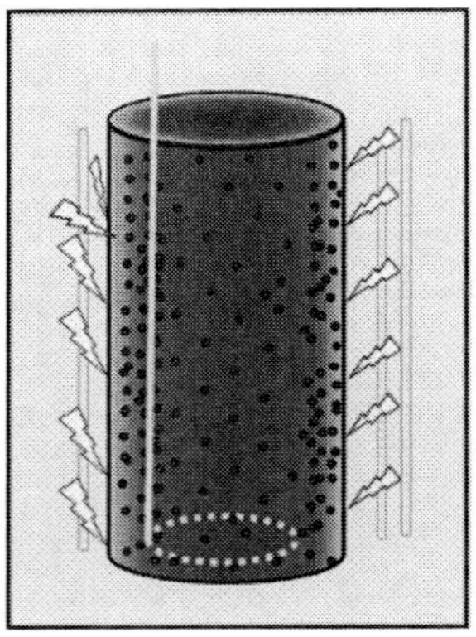

Algal Photobioreactor facilitating axial flow of air only

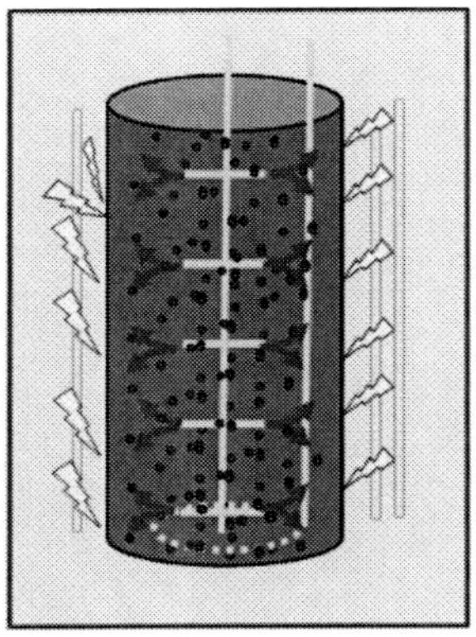

Radially well-mixed Algal Photobioreactor

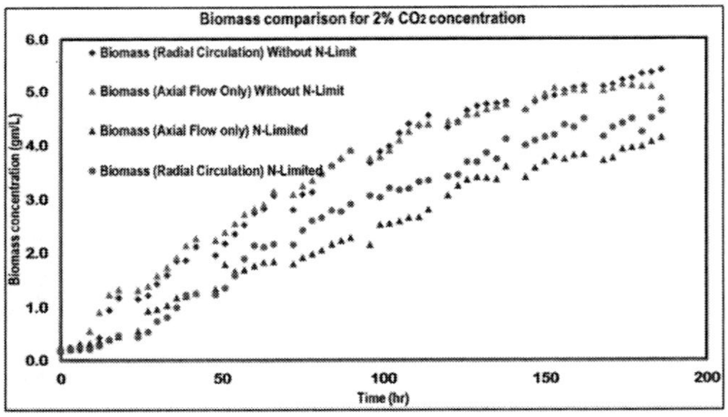

Algal biomass growth dynamics for both cases of reactor configuration

Figure: Microalgal photobioreactor configurations and its effect on algal growth dynamics

139e Systems Design Approach for Dynamic Optimization of Algae Based Ccus Process

Swaminathan Sundar, Rahul Kakodkar, Dustin Kenefake, Efstratios Pistikopoulos

Carbon neutrality has become one of the key initiatives that both governments and industries are striving to achieve. Currently, the energy sector is one of the biggest sources of bulk carbon generation and thus decarbonizing the energy sector is pivotal to achieve net carbon neutrality. To this end, various carbon capture and utilization (CCUS) technologies have been retrofitted in extant processes to capture and convert CO_2 emissions into valuable products [1]. Algae based CCUS process is a novel technology that has shown great promise to potentially feature in future energy systems. Algae as a bio-fuel source has several advantages compared to other traditional plant based biomass, including- i) high biomass growth rate ii) presence of strains that are relatively agnostic to water source type i.e. brackish, freshwater etc. iii) does not directly compete with the FEWN for arable land and fresh water. However, questions do arise on the economical and commercial viability of the algae production process. Harvesting of algae is one of the most energy intensive processes accounting for ~70% of total energy consumption. Optimizing the harvesting schedule is key to maximizing the productivity of algae as high optical density can result in algae cell death which is economically unfavorable. This begs the need for high fidelity process modeling [2] that captures the dynamics of the process and helps in optimizing the process conditions.

In this presentation, we present our progress towards developing a dynamic model to ameliorate the technical feasibility of the process, and identify the optimal process conditions to maximize the production of high value products. For novel CCUS technologies to reach the required technological maturity level for deployment at scale, it is necessary to optimize the process conditions that maximize the economic viability of the process. To this end, development of high fidelity process models that accurately capture the dynamics of the process are necessary. These models can be highly non-linear in nature and accurately predict the dynamics of the system under different process conditions. The algae based CCUS process is an ordinary differential equation (ODE) model that incorporates the *Monod* equations for describing reaction kinetics. The model is dependent on various factors including the concentration of nutrients and CO_2 in the medium, light intensity and factors that are inherent to the strain of the algae. The process is modeled using the pyomo.dae toolbox [3,4], and optimized using a nonlinear programming solver.

References

[1] J. F. D. Tapia, J.-Y. Lee, R. E. Ooi, D. C. Foo, and R. R. Tan, "A review of optimization and decision-making models for the planning of co2 capture, utilization and storage (ccus) systems," Sustainable Production and Consumption, vol. 13, pp. 1–15, 2018.

[2] E. Lee, M. Jalalizadeh, and Q. Zhang, "Growth kinetic models for microalgae cultivation: A review," Algal research, vol. 12, pp. 497–512, 2015.

[3] M. L. Bynum, G. A. Hackebeil, W. E. Hart, C. D. Laird, B. L. Nicholson, J. D. Siirola, J.-P. Watson, and D. L. Woodruff, Pyomo–optimization modeling in python, vol. 67. Springer Science & Business Media, third ed., 2021.

[4] W. E. Hart, J.-P. Watson, and D. L. Woodruff, "Pyomo: modeling and solving mathematical programs in python," Mathematical Programming Computation, vol. 3, no. 3, pp. 219–260, 2011.

139f Amine-Assisted Simultaneous CO_2 Absorption and Mineral Carbonation: Effect of Different Categories of Amines

Yafei Mao

Greenhouse effects caused by the daily increase of CO_2 are becoming more serious recently. The emerging hybrid technology – CO_2 capture & mineral carbonation – provides a novel and effective route in Carbon Capture, Utilization & Storage (CCUS). In this way, CO_2 not only can be absorbed, also can be utilized to produce carbonate product. However, there is a lack of systematic investigation on the amine selection in this hybrid technology. The product properties such as crystal morphology and transformation are not systematically reported. In this work, four amine categories have been selected: primary amine (ethanolamine, MEA), secondary amine (diisopropanolamine, DIPA), tertiary amine (diethylethanolamine, DEAE) and triamine (diethylenetriamine, DETA). Different temperatures and amine concentrations have been studied to comprehensively discuss the carbonation process and product properties. DETA exhibited the highest absorption rate and largest equilibrium CO_2 loading (1.4:1 mol CO_2 / mol amine). DEAE had a ratio of 1:1; while MEA and DIPA performed a similar absorption rate and loading (0.5:1). The carbonation of all amines was enhanced with the increase of temperature except for DEAE. When the amine concentration increased, MEA, DIPA and DETA were limited in the carbonation process, while DEAE showed the fastest carbonation (nearly 100% Ca conversion) with all the investigated concentrations. As regards to crystal transformation, all the formed vaterite partially or totally transformed to calcite over time, except in the case of DETA. Hence, tertiary amines (i.e. DEAE) have a huge potential to enlarge this hybrid technology in the industry.

201a Modeling Reactor Systems for Waste Plastic Pyrolysis

Pratyush Agarwal, David Tremblay, Shu Wang

About 400 million tonnes of plastic waste is generated every year, with most of it being landfilled, incinerated, or mismanaged as litter.[1] Chemical recycling presents a major opportunity to recover and reuse this plastic waste to manufacture new energy and chemical products. In this study, a first-principles kinetic modeling system in Aspen Plus was utilized to model the pyrolysis of a variety of waste plastic feedstocks in a multiphasic reactor. The kinetic model utilizes the method of moments to track the free radical mechanism for polymer decomposition. It accounts for the detailed random scission, β-scission, cyclic scission, depolymerization, H-abstraction, hydrogen shift, debranching, combination, and disproportionation reactions. The reactor was modeled as a screw extruder with vents for the escaping product gas. To model the vapor-liquid phase evolution, the PC-SAFT property model was used to calculate the state of the polymer system, including explicitly tracked low molecular weight oligomers in the system. Additional reactor dynamics and secondary reactions such as char formation can also be included in the model to accurately represent the operation of a real system. The final model can be utilized as a starting point for design and optimization of a waste plastic chemical recycling system in the circular economy paradigm.

[1] United Nations Environment Program. *Visual feature: Beat plastic pollution*. Retrieved March 20, 2023, from https://www.unep.org/interactives/beat-plastic-pollution/

201b Predictive Machine Learning for Thermal Depolymerization of Plastics and Application to Economic Analysis of Municipal Recycling Facilities

Elizabeth Belden, Michael T. Timko

Chemical recycling via thermal processes such as pyrolysis and hydrothermal liquefaction (HTL) is a potentially viable way to convert mixed streams of waste plastics into usable fuels. Unfortunately, experimentally determining product yields for real waste streams can be time and cost prohibitive, especially since the yields are very sensitive to both feed composition and reaction conditions. Data driven methods capable of predicting yields allow for rapid analysis of a specific feedstock's potential for chemical recycling. In previous work[1] a machine learned model capable of predicting oil yields from pyrolysis reactions was developed and used to demonstrate the thermodynamic potential of such a process with real waste. However, this model had a few key limitations, including being applicable only to pyrolysis reactions, and not HTL; restriction to the big six polymers (high- and low-density polyethylene, polypropylene, polystyrene, polyethylene terephthalate and polyvinyl chloride); and not being able to handle reaction/residence time as a model feature. This work aims to address these limitations and develop a more widely applicable model capable of predicting reaction oil yields for both pyrolysis and HTL reactions. The data set used to develop the previous model was expanded to include HTL data without adding more features by converting the polymer classification from mixture composition of the big six polymers to elemental composition of the feed and molecular weight of the plastics monomers contained within it. The inclusion of the molecular weight of the monomer was necessary for differentiating polymers such as polyethylene and polypropylene that have the same elemental composition. This change in classification allows for the inclusion of polymers outside the big six without adding many new features which would risk overfitting while increasing the data set size, particularly for HTL reactions. The updated data set includes 446 data points (increased from 325 previously). This data set also includes reaction time and residence time as features, by allowing both to exist regardless of reactor type. The dataset was used to train and optimize a Random Forest model, which resulted in a test set mean absolute error (MAE) of $\pm$ 11.5%. The new model can be used for a wide range of plastics and is applicable to both pyrolysis and HTL depolymerization technologies. Here, the model was used to predict the oil yield possible from plastic wastes reported from Municipal Recycling Facilities (MRFs) from a variety of locations. These predictions were used to perform a techno-economic analysis (TEA) for conversion of plastic waste to pyrolysis oil. Data driven models such as these are important decision-making tools allowing for the assessment of new feedstocks and the allocation of finite resources.

1. E. R. Belden, M. Rando, O. Ferrara, E. Himebaugh, C. Skangos, N. K. Kazantzis, R.C. Paffenroth, M.T. Timko. Machine learning predictions of oil yields obtained by plastic pyrolysis and application to thermodynamic analysis. *ACS Engineering Au*, 2022

201c Selective Deconstruction and Upcycling of Condensation Polymers

Md Arifuzzaman, Jackie Zheng, Anisur Rahman, Jeffrey Foster, Bobby Sumpter, Tomonori Saito

Condensation polymers such as poly(ethylene terephthalate) (PET), polyamides (PA), polyurethanes (PU), and polycarbonate (PC) comprise ~30% of the global plastic production. The catalytic deconstruction is one of the major paths for chemical recycling of condensation polymers. Although there has been progress on chemical recycling of condensation polymers especially PET, most condensation polymers are not recycled because of the difficulty in depolymerization to pure building blocks especially from mixed state in an energy efficient manner. Here, we have developed a tailored organocatalyst to enable low energy depolymerization pathways for condensation polymers. Our catalyst allows glycolysis of PET, PA, PU, PC and their multiple mixture at moderate temperature with high yield. A wide range of post-consumer plastics waste, such as bottles, packaging, foam, carpet, etc. is readily deconstructed into monomers with high efficiency. The Life Cycle Assessment indicates that the reproduction of various condensation polymers from the deconstructed monomers will result in a significant reduction in greenhouse gas emissions and energy input. Furthermore, we have developed a path to deconstruct those condensation polymers to selective length of oligomers. We have utilized those deconstructed building blocks to synthesize upcycled polymers, and the upcycled polymers can be further deconstructed to reusable building blocks. Such circular design contributes to establishing new closed-loop circularity of polymers by energy efficient selective deconstruction and upcycling of various commodity plastics.

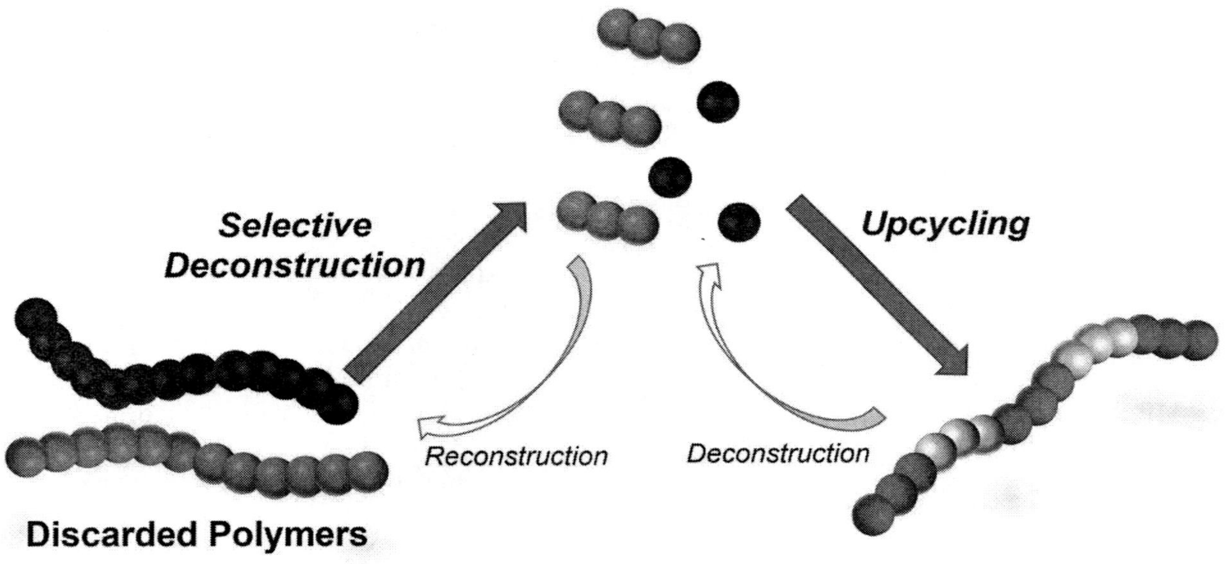

201d Combined Denitrogenation and Olefins Saturation in Hydroprocessing for Chemical Recycling

César Pernalete, Klaus Jacobs, Kevin Van Geem, Joris W. Thybaut

Chemical recycling of plastics via 'conversion' refers to the use of chemical transformations to convert plastic waste into an oil-like hydrocarbon stream. In conversion schemes via thermal pyrolysis the product oil typically has a high olefins content (~50 wt%) [1,2,3]. Further processing of this pyrolysis oil in the refining/petrochemical industry, especially when co-feeding pyrolysis oil with fossil streams, requires a previous hydrodenitrogenation step to guarantee product quality and protect catalyst in downstream hydrocracking processes [4,5,6].

To acquire a deeper understanding of the combined hydrodenitrogenation-olefin saturation, an experimental investigation of the hydrotreating of a mixture containing hexadecene, octadecene and quinoline has been performed. A commercially available $NiMoP/Al_2O_3$ catalyst was employed in a three-phase Robinson Mahoney reactor at a H_2/HC ratio of 890 Nm^3/m^3 and total pressure of 90 bar while the temperature ranged between 300 and 350 °C and the LHSV varied between 0.4 and 1 h^{-1}.

In general, as shown in Figure 1, it was observed that olefin conversion decreased dramatically with LHSV at 300 °C being the lowest between 70% and 75% at the highest LHSV. The higher the temperature the less significant the impact of the LHSV on the conversion was, given the high conversion that was achieved, i.e., at 350 °C, the olefin conversion always exceeded 95%. Quinoline conversion was always higher than 95%, the maximum yield of denitrogenated products amounting to 15% only. The hydrogenated intermediate 1,2,3,4 tetrahydroquinoline (14THQ) was identified as the major product in all cases.

Figure 1. Olefins conversion vs. LHSV (h^{-1}) at different operating temperatures and 90 Bar

By comparing these results with similar studies found in literature it can be deduced that olefin saturation interferes with hydrogenation and ring opening reactions of the intermediates of quinoline hydrodenitrogenation as lower nitrogen conversions than in the literature at similar operating conditions were obtained [7,8,9,10,11].

As a future work intrinsic kinetic parameters for olefin hydrogenation obtained from this experimental campaign are intended to be used to tune olefins hydrogenation kinetics in an industrial scale reactor model for hydroprocessing (hydrotreating + hydrocracking) configurations where the reactor is co-fed with a fossil-pyrolysis oil mixture [12].

Keywords: chemical recycling, pyrolysis oil, olefin saturation, quinoline hydrodenitrogenation, hydrotreating

References

[1] D. Almeida and M. de F. Marques, "Thermal and catalytic pyrolysis of plastic waste," Polímeros, vol. 26, pp. 44–51, Mar. 2016, doi: 10.1590/0104-1428.2100.

[2] M. Kusenberg et al., "A comprehensive experimental investigation of plastic waste pyrolysis oil quality and its dependence on the plastic waste composition," Fuel Processing Technology, vol. 227, p. 107090, Mar. 2022, doi: 10.1016/j.fuproc.2021.107090.

[3] H. Dao Thi, M. R. Djokic, and K. M. Van Geem, "Detailed Group-Type Characterization of Plastic-Waste Pyrolysis Oils: By Comprehensive Two-Dimensional Gas Chromatography Including Linear, Branched, and Di-Olefins," Separations, vol. 8, no. 7, Art. no. 7, Jul. 2021, doi: 10.3390/separations8070103.

[4] C. S. Raghuveer, J. W. Thybaut, R. De Bruycker, K. Metaxas, T. Bera, and G. B. Marin, "Pyridine hydrodenitrogenation over industrial NiMo/γ-Al2O3 catalyst: Application of gas phase kinetic models to liquid phase reactions," Fuel, vol. 125, pp. 206–218, Jun. 2014, doi: 10.1016/j.fuel.2014.02.017.

[5] Q. Xin, A. Alvarez-Majmutov, H. D. Dettman, and J. Chen, "Hydrogenation of Olefins in Bitumen-Derived Naphtha over a Commercial Hydrotreating Catalyst," Energy Fuels, vol. 32, no. 5, pp. 6167–6175, May 2018, doi: 10.1021/acs.energyfuels.8b00344.

[6] Q. Xin, A. Alvarez-Majmutov, R. Gieleciak, J. Chen, and H. Dettman, "Hydrotreating of Olefins in Thermally Processed Bitumen under Mild Conditions," Energy & Fuels, vol. 33, Mar. 2019, doi: 10.1021/acs.energyfuels.9b00157.

[7] "Hydrodenitrogenation of quinoline and its intermediates over sulfided NiW/γ-Al2O3 in the absence and presence of H2S - Luan - 2009 - Asia-Pacific Journal of Chemical Engineering - Wiley Online Library." https://onlinelibrary.wiley.com/doi/10.1002/apj.322 (accessed Mar. 31, 2022).

[8] C. Tu, S. Liu, C. Liu, X. Wu, Q. Chen, and W. Huang, "Single-Pot Synthesis of NiMo-Al2O3 Catalyst for Quinoline Hydrodenitrogenation," Energy Fuels, vol. 35, no. 2, pp. 1120–1128, Jan. 2021, doi: 10.1021/acs.energyfuels.0c03186.

[9] S. Tian, X. Li, A. Wang, Y. Chen, H. Li, and Y. Hu, "Hydrodenitrogenation of Quinoline and Decahydroquinoline Over a Surface Nickel Phosphosulfide Phase," Catal Lett, vol. 148, no. 6, pp. 1579–1588, Jun. 2018, doi: 10.1007/s10562-018-2370-z.

[10] M.-T. Nguyen, M. Tayakout-Fayolle, G. D. Pirngruber, F. Chainet, and C. Geantet, "Kinetic Modeling of Quinoline Hydrodenitrogenation over a NiMo(P)/Al2O3 Catalyst in a Batch Reactor," Ind. Eng. Chem. Res., vol. 54, no. 38, pp. 9278–9288, Sep. 2015, doi: 10.1021/acs.iecr.5b02175.

[11] B. Mattson et al., "Heterogeneous Catalysis: The Horiuti-Polanyi Mechanism and Alkene Hydrogenation," Journal of Chemical Education, vol. 90, pp. 613–619, May 2013, doi: 10.1021/ed300437k.

[12] Pernalete C., Ibanez J., Van Geem K., Thybaut J., 'Hydrocracking of complex mixtures: From bulk properties, over fundamental kinetics to detailed product composition,' Catalysis Today, vol. 378 (2021) pp. 189–201. doi: 10.1016/j.cattod.2021.06.010.

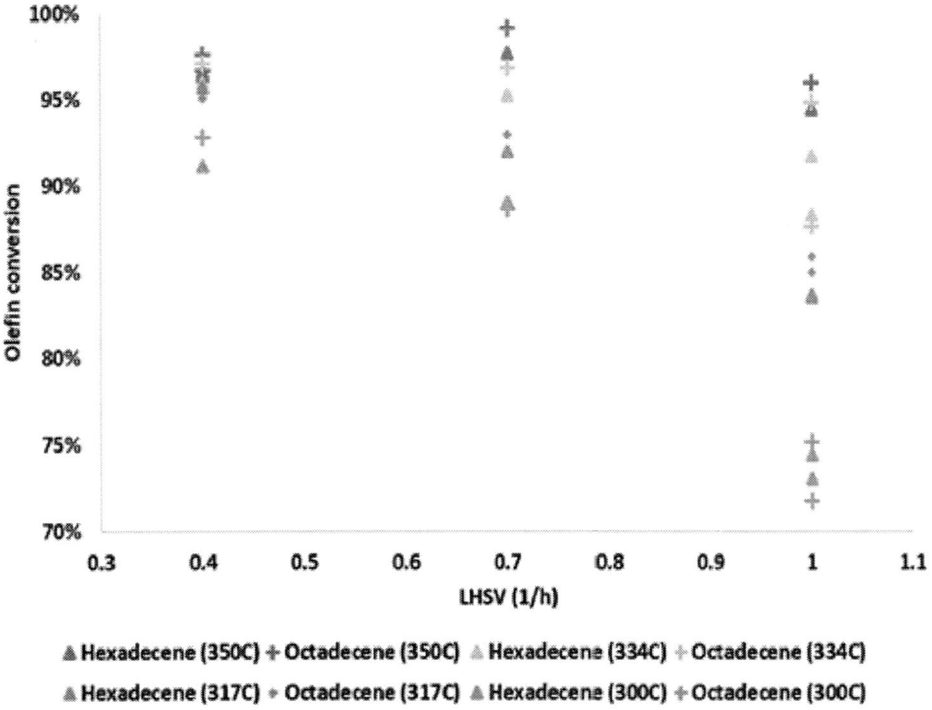
100%
95%
90%
85%
80%
75%
70%
Olefin conversion
0.3
0.4
0.5
0.6
0.7
0.8
0.9
1
1.1
LHSV (1/h)
Hexadecene (350C)
Octadecene (350C)
Hexadecene (334C)
Octadecene (334C)
Hexadecene (317C)
Octadecene (317C)
Hexadecene (300C)
Octadecene (300C)

201e Plastic Waste Pyrolysis Wax Production, Characterization, and Application

Ali Zolghadr, Arya Lekshmi Anilkumar, Daniel Kulas, Andrew Walch, David Shonnard

The complex challenge of recycling plastics can be enabled through innovations in chemical recycling, an example of which is thermochemical conversion processes such as pyrolysis. Fast pyrolysis is considered an effective method for transforming certain types of plastic waste into valuable products that can be upcycled into new plastics or high value chemicals. Polyethylene (PE), the most commonly used plastic globally with a consumption share of approximately 30%, can also be processed through such methods to yield useful products. However, one of the major challenges is ensuring the consistent and efficient production of high-quality, valuable products from transformed plastic waste. One promising high-quality product from pyrolysis is industrial wax, which can be used as a feedstock for many items including candles, crayons, adhesives, and cosmetics.

This study employed a novel liquid-fed waste plastic fast-pyrolysis pilot plant equipped with a vertical pyrolysis reactor, designed to effectively collect char (char chamber), to process low-density polyethylene (LDPE) from military Meals, Ready-to-Eat (MRE) bags and high-density polyethylene (HDPE) from MRE bags. The HDPE and LDPE bags are formulated with about 3% wt. nanoclay as a strengthening agent, which is the source of char from the process. The waste plastic feeds were combined with recycled pyrolysis wax in a 1:1 ratio (wt.) to form a liquid feed for pyrolysis. The fast pyrolysis reaction occurred at 460°C at vapor residence times less than 1 second, resulting in a wax yield rate of >70% and 100% conversion of the waste feed into gas, liquid, wax, and char products. The resulting pyrolysis wax underwent evaporation to remove light species and was hydrotreated using hydrogen gas and Pt/Al_2O_3 catalyst. The pyrolysis wax from each step and hydrotreated products were characterized via thermogravimetric analysis (TGA), Fourier-transform infrared spectroscopy (FTIR), X-ray diffraction (XRD), and gas chromatography-mass spectrometry (GC-MS) analysis. According to the TGA and GC-MS results, the hydrocarbon number ranges in pyrolysis wax became more similar to paraffin wax by evaporating 25% of the volatile species (<C20) from the crude wax sample. Interestingly, the alkene peaks in the GC-MS spectra decreased compared to the "shoulder" alkane peaks, disappearing as the molecular weight of the sample increased (>C20), with emergence of the alkane as confirmed by FTIR results. Additionally, the mild hydrotreatment successfully converted the alkene compounds to alkane within 2 hours, creating a suitable industrial wax alternative. These findings demonstrate the potential of pyrolysis wax as a promising alternative to paraffin wax in industrial wax applications and for candle-making purposes.

201f Superacid-Assisted Polystyrene Upcycling

Atharwa Thigale, Carrie L. Trant, Juhong Ahn, Melissa Nelson, Chang Yeol Ryu, Chulsung Bae, Sangwoo Lee

Amongst the recycling technologies for end-of-life polystyrene (PS) products, chemical upcycling is a promising way to build a profitable circular plastic economy for PS. Acid-catalyzed conversion of PS requires a relatively milder reaction environment and has the potential to produce high-value chemical compounds in a low-cost environment. We investigated the superacid-assisted degradation of PS chains at 20 °C using time-resolved characterizations to understand how the PS chains transform into small compounds. The following stages are identified: i) topological changes in the PS chains by *intra*-chain crosslinking that forms nanoparticles and simultaneous transformation of styrene to indan units, ii) breakdown of the crosslinked polymer nanoparticles into a range of small compounds such as benzene and indan/indene derivatives, and iii) and formation of the polycyclic aromatic hydrocarbons by the Scholl reaction. Our findings elucidate the precise degradation process of PS and will help design effective and economically competitive upcycling strategies for end-of-life polymers.

201g Novel Thin Film Composite Membranes Fabricated from Upcycled Waste PVC Pipe Support Layer for Water Treatment

Atta Ur Razzaq, Milad Esfahani

Nanofiltration (NF) thin-film-composite (TFC) membranes are well-known energy-efficient membranes for water treatment. Generally, these membranes are fabricated on support layer made of commercially available materials such as polyethersulfone (PES). On the other hand, plastic pollution is one of the main environmental concerns. Therefore, it is beneficial to upcycle waste plastic to a high-added-value product such as a membrane. In this work, we demonstrate the successful fabrication of the thin film composite (TFC) membranes with the support layer made from upcycled waste PVC pipe. The ultrafiltration (UF) support membranes were fabricated from waste PVC pipe (22 wt.% polymer in a mixture of N,N-Dimethylformamide and Tetrahydrofuran 3:2 (v/v)) via nonsolvent induced phase separation (NIPS) and then a polyamide layer was synthesized on the upcycled PVC membrane using interfacial polymerization. The physicochemical properties of the upcycled PVC membrane were compared and benchmarked with membranes fabricated using research-grade research (RG) PVC and commercial PES support layer. The results indicated that upcycled PVC pipe TFC membrane exhibited similar hydrophilicity (water contact angle around 30°), slightly more negative surface charge, and higher water permeability at 100 psi (17.2 LMH/bar) compared to RG TFC (14.5 LMH/bar) and PES TFC membranes (12.7 LMH/bar). Salt rejection of waste PVC pipe TFC membrane was also comparable to RG TFC and PES TFC as it achieved 88.9 % rejection for $MgSO_4$, 77.8 % rejection for $CaCl_2$, and 46.9% rejection for NaCl. Since the upcycled waste PVC pipe TFC membranes showed similar physiochemical properties and performance to the commercial RG PVC and PES TFC membranes, this shows that upcycling waste PVC pipe to membranes for water treatment can solve two biggest global challenges of plastic pollution and water treatment/purification simultaneously.

253a Investigating and Leveraging Natural Variation of Lignin in Biomass

Arthur Ragauskas

Over the past decade, the biorefining of biomass has evolved from solely focusing on renewable fuel generation to the complete utilization of biomass for fuels, chemicals, and materials. As such, the development of lignin first pretreatments has been pursued to yield a clean stream of lignin and a reactive, low-recalcitrance cellulose-rich stream. The former stream provides a bioresource for new materials and fuel opportunities, whereas the latter provides enhanced conversion to cellulosic ethanol. The continued interest in renewable fuels from biomass has placed a particular interest in biomass fractionation into streams enriched in cellulose, hemicellulose, and lignin. In this presentation, the natural variation of cell wall polymers is explored in Populus and switchgrass, and we examine how advanced lignin first pretreatments can be tailored to enhance fractionation and conversion by utilizing recent advances in spectroscopic analysis and computational modeling.This presentation will examine the unique chemistry interactions between pretreatment solvents and cell wall components and their contribution to reducing biomass recalcitrance and efficient conversion.

253b Maleic Acid Hydrotropic Lignin (MAHL) for Efficient Conversion to Lignin Aromatics: Production, Characterization, Conversion

Junyong Zhu

In this presentation, I will describe maleic acid hydrotropic fractionation (MAHF) of wood for producing highly reactive lignin for efficient conversion to lignin aromatics. MAHF can achieve rapid (~ 30 min) dissolution of wood lignin at ~100°C at high yield in aqueous systems. The dissolved lignin can be easily separated through precipitation by simply diluting the spent liquor with water to the minimal hydrotrope concentration (MHC). The dissolved xylose can be efficiently dehydrated into furfural, a high value chemical, using the acid hydrotrope (maleic acid) in the spent liquor without additional catalyst. The rapid solubilization of lignin at low temperatures resulted in lignin with low degree of condensation and highly reactive, which facilitate catalytic conversion to lignin aromatics. We will compare the properties, especially reactivity and conversion yields of MA hydrotropic lignin (MAHL) from batch reaction with those from flow through reaction. Both catalytic oxidative depolymerization and supercritical methanol dehydration hydrodeoxygenation processes were applied to MAHL from batch and flow through reactions. The results show MAHL can be efficiently converted to lignin monophenols. MAHL from flow through reaction are even more reactive with higher conversion yield due to the high content beta-O-4 linkages retained. MA is a US FDA approved indirect food additive, the low solubility of MA also facilitates MA recovery, therefore MAHF has significant advantages over most existing fractionation processes especially those solvent-based processes.

253c Predicting and Improving Lignin Depolymerization Processes with Whole Plant Cell Wall $HSQC_0$ NMR and Kinetic Monte Carlo Simulation

Claire Bourmaud, Qiang Li, Yifan Wang, Victor Sabanza Gil, Songlan Sun, Stefania Bertella, Gregory G. Facas, Gregg Beckham, Philippe Schwaller, Dionisios Vlachos, Jeremy Luterbacher

Lignin depolymerization via hydrogenolysis is a promising pathway to produce renewable phenolic monomers.[1] However, our understanding of this process is limited by lignin heterogeneous structure, following complex deconstruction mechanism and yielding broad product distribution after β-O-4 ether linkages.[2] We developed a rapid extrapolated time zero ^1H–^{13}C heteronuclear single quantum coherence nuclear magnetic resonance ($HSQC_0$ NMR) sequence that enables to quantify native lignin structural features with whole plant cell wall (WPCW) NMR spectroscopy, overcoming fast spin relaxation in gel phase.[3,4] This analytical method is a promising tool to gain new insights towards native lignin structure.

Kinetic and mechanistic studies of β-O-4 model compounds depolymerization were then performed and confirmed higher reactivity of lignin end-units when adsorbed on a Ru/C catalyst. The obtained reaction network and NMR structural insights were applied to multiscale *LigninGraph* libraries via a kinetic Monte Carlo framework to model lignin hydrogenolysis.[5,6] Reductive Catalytic Fractionation was successfully modelled and first-ever predicted by our algorithm using dimer kinetic data. This model can notably establish structure-reactivity relationships, that expands our knowledge of lignin structure and understanding of extraction processes. Reaching unexpected insights for lignin engineering, we could ultimately guide the design of highly reactive lignin moieties towards hydrogenolysis.

References

1. Mahdi M. Abu-Omar, Katalin Barta, Gregg T. Beckham, Jeremy S. Luterbacher, John Ralph, Roberto Rinaldi, Yuriy Román-Leshkov, Joseph S. M. Samec, Bert F. Sels, and Feng Wang. "Guidelines for Performing Lignin-First Biorefining." *Energy & Environmental Science,* **2021**, 14, 1, 262–92.

2. John Ralph, Catherine Lapierre, and Wout Boerjan. "Lignin Structure and Its Engineering." *Current Opinion in Biotechnology,* **2019**, 56, 240–49.

3. Masoud Talebi Amiri, Stefania Bertella, Ydna M. Questell-Santiago, and Jeremy S. Luterbacher. "Establishing Lignin Structure-Upgradeability Relationships Using Quantitative ^1H–^{13}C Heteronuclear Single Quantum Coherence Nuclear Magnetic Resonance (HSQC-NMR) Spectroscopy." *Chemical Science,* **2019**, 10, 35, 8135–42.

4. Shawn D. Mansfield, Hoon Kim, Fachuang Lu, and John Ralph. "Whole Plant Cell Wall Characterization Using Solution-State 2D NMR." *Nature Protocols* **2012**, 7, 9, 1579–89.

5. Yifan Wang, Jake Kalscheur, Elvis Ebikade, Qiang Li, and Dionisios G. Vlachos. "LigninGraphs: Lignin Structure Determination with Multiscale Graph Modeling." *Journal of Cheminformatics* **2022**, 14, 1, 43.

6. R. Vinu, Seth E. Levine, Lin Wang, and Linda J. Broadbelt. “Detailed Mechanistic Modeling of Poly(Styrene Peroxide) Pyrolysis Using Kinetic Monte Carlo Simulation.” *Chemical Engineering Science,* **2012**, 69, 1, 456–71.

253d Lignin Degradation Mechanisms of *Rhodopseudomonas Palustris* CGA009

Mark Kathol, Niaz Chowdhury, Cheryl Immethun, Rajib Saha

Lignin is a complex and abundant biopolymer that constitutes a significant proportion of the structural material of many plants. However, due to its lack of useful chemistry and recalcitrance as a carbon feedstock for microorganisms, it is commonly treated as waste in industries where it is a byproduct, such as in paper manufacturing and ethanol production from corn stover hydrolysate. Lignin requires complex biochemistry for its degradation, which is typically supplied from a community of microorganisms and fungi, as a result, an overwhelming majority of annually produced lignin is instead burned as fuel. With the pollutants and non-degradable products from abiotic chemical synthesis having an ever more pronounced effect on the environment, there remains a large potential for lignin to be valorized into more useful chemicals which these abiotic processes manufacture. *Rhodopsuedomonas palustris* (*R. palustris*), a non-model, gram-negative soil bacterium with a wide array of unique metabolic features, has demonstrated its ability to catabolize many lignin monomers, and with its unique biochemistry that allows it to produce many sought-after chemicals such as bioplastic and biofuels, it has the potential to become a metabolic engineering chassis. From analysis of growth data collected in this study, *R. palustris* is able to aerobically and anaerobically catabolize multiple lignin breakdown products (LBPs), including p-coumarate, sodium ferulate, p-coumaryl, coniferyl, and sinapyl alcohols. In this study, we investigate the lignin catabolic pathways of *R. palustris* through a "multi-omics" perspective by combining both proteomic and transcriptomic profiles of *R. palustris* cultured with several LBPs. Through initial analysis of transcriptomic data, potential pathways for the aerobic catabolism of each LBP have been identified, while anaerobically, p-coumarate appears to be upregulated in all cases. This study will advance the understanding of the complex carbon metabolism of *R. palustris*, and assist in enabling sustainable biochemical production.

Brown, B., Wilkins, M., & Saha, R. (2022). Rhodopseudomonas palustris: A biotechnology chassis. In *Biotechnology Advances* (Vol. 60). Elsevier Inc. https://doi.org/10.1016/j.biotechadv.2022.108001

Immethun, C. M., Kathol, M., Changa, T., & Saha, R. (2022). Synthetic Biology Tool Development Advances Predictable Gene Expression in the Metabolically Versatile Soil Bacterium Rhodopseudomonas palustris. *Frontiers in Bioengineering and Biotechnology, 10.* https://doi.org/10.3389/fbioe.2022.800734

Morris, D. S., Chowdhury, N. B., Kathol, M., & Saha, R. (2023). Opening the door to nitrogen fixation in oxygenic phototrophs. In *Trends in Biotechnology* (Vol. 41, Issue 3, pp. 283–285). Elsevier Ltd. https://doi.org/10.1016/j.tibtech.2023.01.005

253e Fractionation and Chemical Modification of Celf Lignin: Toward Production of Next-Generation of Polyurethane

Xianzhi Meng, Yunxuan Wang, Zhaohong Xiu, Brent Scheidemantle, Charles M. Cai, Chang Geun Yoo, Arthur Ragauskas

The effective valorization of lignin to renewable polymers such as polyurethanes (PU) is a promising pathway for producing renewable plastics, provided that the process can be realized cost-effectively using environmentally sustainable methods (e.g., isocyanate free). A prerequisite of utilizing lignin in such area is to develop efficient lignin fractionation and modification strategies that can decrease the lignin heterogeneity and tune the chemical structure of lignin streams. For this purpose, several common organic solvents were screened for their ability to fractionate the Co-Solvent Enhanced Lignocellulosic Fractionation (CELF) lignin. A solvent extraction sequence is then proposed to obtain CELF lignin fractions with different chemical functionalities and molecular weights. The obtained CELF lignin fractions were subsequently modified by epoxidation followed by CO_2 cycloaddition to produce carbonated CELF lignin. A green and non-toxic CELF lignin-based non-isocyanate polyurethanes (NIPUs) were finally synthesized via the polyaddition between the tailored carbonated CELF lignin and commercially available diamines. The final NIPU products obtained using this synthetic route were analyzed for their morphological, structural, and thermal properties, and the relationship between lignin structures and NIPU properties was established.

253f Conversion of Corn Stover to Fuel and Chemical – the Optimization of Biomass Carbon Efficiency through Lignin Utilization

Pimphan Meyer, Lesley J. Snowden-Swan, Kyle Pomraning, Steven Phillips

On a dry mass basis, cellulosic feedstock like corn stover typically comprises about 60% carbohydrate and 20% lignin. Typically, a bioconversion method is applied to convert the carbohydrate to value-added chemicals or biofuel. The simplest option to utilize the lignin effluent is generating heat and power onsite; however, this does not optimize either the overall biomass carbon efficiency or the production cost. Process models of hydrothermal liquefaction (HTL) of the leftover lignin from biochemical process were developed based on experimental data and a yield predictive model. Mass and energy balances of the model simulation suggests that producing marketable chemicals from carbohydrate and producing transportation fuel from lignin is probably the most efficient practice to maximize corn stover carbon efficiency. Techno-economic analysis (TEA) is applied to identify significant process cost drivers. Data gaps in the analysis will be addressed.

253g Lignin Derived Ionic Liquids: Synthesis and Application for Biopolymer Processing

Ning Sun, Chang Dou, Aaron Socha

Certain types of ionic liquids (ILs) have shown great solvation power for biopolymers, resulting in remarkable biomass pretreatment efficiency as well as novel approaches to produce biocomposites. To date, the best performing IL cations such as imidazolium and cholinium, are derived from petroleum and/or natural gas. ILs derived from inexpensive and renewable sources are thus highly desirable. Lignin is Earth's most abundant, naturally-occurring aromatic polymer, and replacement of petroleum-based platforms with those derived from lignin represents a virtuous challenge for chemists. The synthesis of novel benzyl alkylammonium ILs from commercial lignin was performed using batch reactor systems followed by facile ion exchange chemistry. Both lignin oxidation and reduction has been studied to produce intermediates molecules for IL synthesis. The oxidation process has been scaled up to 10 L batch reactor followed by product recovery through filtration. Approximately 90% yields were achieved over three synthetic steps, and all product materials were fully characterized. The structure activity relationships of the lignin-derived ILs were evaluated for their ability to dissolve microcrystalline cellulose. A technoeconomic analysis and life cycle analysis has been performed to evaluate the economic viability and carbon footprint of the lignin-derived ionic liquid synthesis process.

253h Hydrodeoxygenation of Bio-Oil Via in-Situ Alkene Dehydrogenation

Abdullah Baubaid, Marcos Millan

The abundance of waste biomass and its extracted lignin and its intrinsic aromatic network could provide the chemical industry with a rich, sustainable, and renewable hydrocarbon supply. This potential source of green added-value chemicals, like benzene, toluene and xylenes (BTX) boosts the interest on biorefinery and its integration with the existing petrochemical state-of-the-art processes. With the relatively high oxygen content in bio-oil, which hinders its heating value and processability, hydrodeoxygenation (HDO) becomes a crucial step in biorefineries [1], [2]. Conventionally, HDO is processed using high-pressure (up to 200 bar) expensive hydrogen [1]. Consequently, the use of a hydrogen donor is investigated as an alternate strategy by using solvent and/or co-feeding at different operating conditions [3]–[5]. Optimally, the reaction used to produce hydrogen should in itself also lead to added-value byproducts. Hence, dehydrogenation of alkanes can be an interesting choice as valuable alkenes would be generated as shown in Figure 1.

This research aims to demonstrate HDO of bio-oil via in-situ alkane dehydrogenation. Alkanes such as ethane and propane are the primary feeds of the petrochemical value chain to produce the building block chemicals ethylene, propylene, and aromatics. Therefore, the incorporation of alkenes production with upgraded bio-oil would improve the overall outlook of HDO. This research seeks to increase bio-oil conversion due to the catalytic formation of reactive hydrogen at the catalyst surface derived from the dehydrogenation of alkenes. In addition, this research seeks to increase alkane conversion derived from the continuous consumption of hydrogen for bio-oil upgrading. To investigate alkane and bio-oil conversions, experiments are conducted on a single catalytic system comprised of HDO and alkane dehydrogenation catalysts in a continuous up-flow reactor. To achieve higher hydrogen formation from thermodynamic equilibrium perspective, these experiments are conducted at atmospheric pressure and high temperature environment using guaiacol as model compound. Catalytically, these studies were conducted on Ni-Mo and MoO_3 based catalyst for HDO, and Cr-based catalyst for alkane dehydrogenation [1], [6]–[9].

A bio-oil conversion similar to that achieved through the use of hydrogen was observed, with products having a higher C/O ratio, as well as ethylene production, when using alkane dehydrogenation catalysts and a superior conversion with HDO catalysts. In addition, increases in reaction temperature and catalyst loading have a significant positive effect on HDO performance. On the other hand, catalytic dehydrogenation of ethane suffers a decline in conversion after 30 minutes of time-on-stream until it reaches a plateau; the introduction of bio-oil model compounds accelerates this decline through catalyst deactivation, necessitating frequent of catalyst re-activation. In summary, the results demonstrated the feasibility of replacing hydrogen with alkanes and there is scope by further developing the process through more effective catalyst formulations.

References

[1] E. Furimsky, “Catalytic hydrodeoxygenation,” *Appl. Catal. A Gen.*, vol. 199, no. 2, pp. 147–190, Jun. 2000, doi: 10.1016/S0926-860X(99)00555-4.

[2] Z. Si, X. Zhang, C. Wang, L. Ma, and R. Dong, “An Overview on Catalytic Hydrodeoxygenation of Pyrolysis Oil and Its Model Compounds,” *Catal. 2017, Vol. 7, Page 169*, vol. 7, no. 6, p. 169, Jun. 2017, doi: 10.3390/CATAL7060169.

[3] A. T. To and D. E. Resasco, "Hydride transfer between a phenolic surface pool and reactant paraffins in the catalytic cracking of m-cresol/hexanes mixtures over an HY zeolite," *J. Catal.*, vol. 329, pp. 57–68, Sep. 2015, doi: 10.1016/J.JCAT.2015.04.025.

[4] M. E. Domine, A. C. Van Veen, Y. Schuurman, and C. Mirodatos, "Coprocessing of Oxygenated Biomass Compounds and Hydrocarbons for the Production of Sustainable Fuel," *ChemSusChem*, vol. 1, no. 3, pp. 179–181, Mar. 2008, doi: 10.1002/CSSC.200700049.

[5] X. dong Fan *et al.*, "Benzene, toluene and xylene (BTX) from in-situ gas phase hydrodeoxygenation of guaiacol with liquid hydrogen donor over bifunctional non-noble-metal zeolite catalysts," *Renew. Energy*, vol. 152, pp. 1391–1402, Jun. 2020, doi: 10.1016/j.renene.2020.01.015.

[6] U. Olsbye, A. Virnovskaia, O. Prytz, S. J. Tinnemans, and B. M. Weckhuysen, "Mechanistic Insight in the Ethane Dehydrogenation Reaction over Cr/Al2O3 Catalysts," *Catal. Lett. 2005 1031*, vol. 103, no. 1, pp. 143–148, Sep. 2005, doi: 10.1007/S10562-005-6521-7.

[7] B. M. Weckhuysen and R. A. Schoonheydt, "Alkane dehydrogenation over supported chromium oxide catalysts," *Catal. Today*, vol. 51, no. 2, pp. 223–232, Jun. 1999, doi: 10.1016/S0920-5861(99)00047-4.

[8] G. Zhou, P. A. Jensen, D. M. Le, N. O. Knudsen, and A. D. Jensen, "Atmospheric Hydrodeoxygenation of Biomass Fast Pyrolysis Vapor by MoO3," *ACS Sustain. Chem. Eng.*, vol. 4, no. 10, pp. 5432–5440, Oct. 2016, doi: 10.1021/acssuschemeng.6b00757.

[9] I. D. Mora-Vergara, L. Hernández Moscoso, E. M. Gaigneaux, S. A. Giraldo, and V. G. Baldovino-Medrano, "Hydrodeoxygenation of guaiacol using NiMo and CoMo catalysts supported on alumina modified with potassium," *Catal. Today*, vol. 302, pp. 125–135, Mar. 2018, doi: 10.1016/j.cattod.2017.07.015.

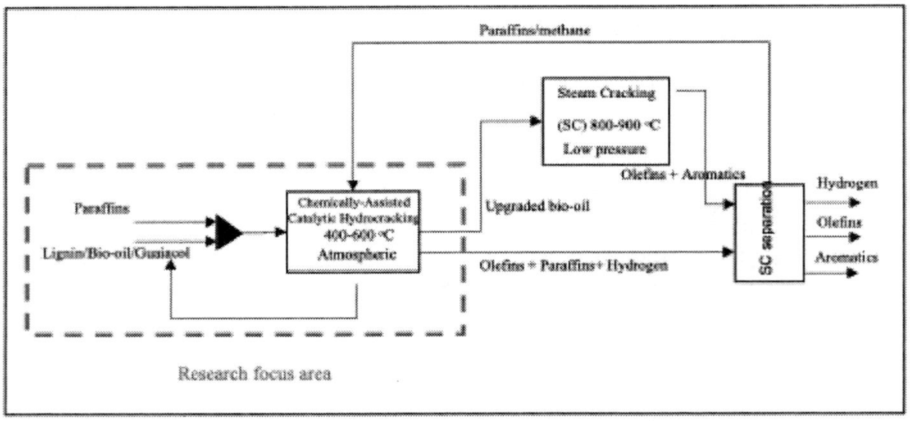

253i Complete Hydrodeoxygenation of Enzymatic Hydrolysis Lignin to Cycloalkane Fuels

Yongdan Li, Yushuai Sang, Hairui Jiao, Hong Chen

Lignocellulose, mainly composed of cellulose, hemicellulose and lignin, is a sustainable feedstock for producing commodity chemicals and fuels. For hundreds of years, cellulose has been utilized in paper production. Recently, cellulose and hemicellulose are converted into bio-ethanol in 2G biorefineries, but enzymatic hydrolysis lignin (EHL) from 2G bioethanol production are poorly utilized, and are regarded as a large volume solid waste [1]. Our previous works focused on EHL depolymerization in ethanol, and produced phenolic products with unsupported Ni [2], Ni-Mo alloy [3] and WO_3 supported on Al_2O_3 [4] as catalysts. Recently, a one-pot method for simultaneous lignin depolymerization and hydrodeoxygenation (HDO) of products is developed. Although noble metals are excellent catalysts for HDO reactions, their high cost hinders their application. In our recent works, we developed non-noble metal catalysts from hydrotalcite precursors, which achieved complete hydrodeoxygenation of EHL into cycloalkane fuel without formation of char.

Ni catalyst derived from the reduction of Ni-Al hydrotalcite (NiAl-LDH) was employed for HDO of EHL. The effects of Ni/Al ratio and reduction condition on the catalyst activity have been examined. The catalyst obtained from the reduction of NiAl-LDH at 460 ^{o}C (Re-NiAl-LDH-460) showed the highest activity among the catalyst samples examined. With this catalyst, EHL was completely converted and depolymerized at 320 °C under 3 MPa H_2 for 6 h, yielding 133.24 mg/g EHL of cycloalkanes. Meanwhile, the heating value of the product is 42.5 MJ/kg and much higher than that obtained without a catalyst (25 MJ/kg) under the same reaction condition. After that, Fe was introduced into the Ni-Al hydrotalcite, followed by reduction to prepare a NiFe alloy catalyst. The total yield of cycloalkanes was improved to 161.1 mg/g EHL under the same reaction conditions (320 °C, 3 MPa H_2, 6 h). The results of the lignin dimer model compound conversion demonstrate that the NiFe alloy catalyst exhibits high activity in breaking C-O linkages and achieves partial cleavage of C-C linkages.

Reference

[1] L. Dessbesell, M. Paleologou, M. Leitch, R. Pulkki, C. Xu, Global lignin supply overview and kraft lignin potential as an alternative for petroleum-based polymers, Renewable & Sustainable Energy Reviews, 123 (2020).

[2] Y. Sang, M. Chen, F. Yan, K. Wu, Y. Bai, Q. Liu, H. Chen, Y. Li, Catalytic Depolymerization of Enzymatic Hydrolysis Lignin into Monomers over an Unsupported Nickel Catalyst in Supercritical Ethanol, Industrial & Engineering Chemistry Research, 59 (2020) 7466-7474.

[3] Y. Bai, K. Cui, Y. Sang, K. Wu, F. Yan, F. Mai, Z. Ma, Z. Wen, H. Chen, M. Chen, Y. Li, Catalytic Depolymerization of a Lignin-Rich Corncob Residue into Aromatics in Supercritical Ethanol over an Alumina-Supported NiMo Alloy Catalyst, Energy & Fuels, 33 (2019) 8657-8665.

[4] F. Mai, Z. Wen, Y. Bai, Z. Ma, K. Cui, K. Wu, F. Yan, H. Chen, Y. Li, Selective Conversion of Enzymatic Hydrolysis Lignin into Alkylphenols in Supercritical Ethanol over a WO3/γ-Al2O3 Catalyst, Industrial & Engineering Chemistry Research, 58 (2019) 10255-10263.

253j Continuous Selective Hydrodeoxygenation (HDO) of Guaiacol to Cycloalkanes Under Near-Ambient Conditions

Adarsh Kumar, Bin Yang, Huamin Wang, Daniel Santosa

Many parallel efforts are underway to produce next-generation clean fuels from lignin in order to reduce reliance on petroleum feedstock. In this work, we developed a continuous selective hydrodeoxygenation (HDO) process of guaiacol conversion to cycloalkanes under near-ambient conditions, leading to deeper understanding of HDO chemistry of lignin conversion. Previous reports of guaiacol conversion under near-ambient conditions (<200 ^{o}C and 4 MPa) suffers from low conversion and low selectivity of HDO products in addition to catalytic site poisoning. In order to overcome these challenges, the typical incipient wetness impregnation method was modified to improve HY zeolite-supported Ru catalyst synthesis. Results showed that metal dispersion and acid site uniformity were improved. The modified catalyst (Ru-HY-60-MI) was then tested in a continuous fixed bed reactor, resulting in increased HDO reaction of guaiacol to cycloalkanes at 180 ^{o}C and 1 MPa compared to the original catalyst. Pressure, temperature and WHSV dependent tests validated guaiacol HDO mechanism over Ru-HY-60-MI through metal-catalyzed ring hydrogenation of guaiacol to 2-methoxy-cyclohexanol followed by acid-catalyzed demethoxylation and dehydration to cyclohexene and subsequent metal-catalyzed hydrogenation of cyclohexene to cyclohexane. This insightful understanding of continuous guaiacol HDO is key to improve the efficiency of the continuous HDO of lignin to jet fuel hydrocarbons and enable the process scale-up for a more sustainable biorefinery.

339a Cost-Effective Methanol Production from Vacuum Residue Gasification

Umer Zahid

The continuous rise of global carbon emissions demands the utilization of fossil fuels in a cleaner and sustainable way. Gasification is a potential technology that can convert dirty fossil fuels into clean and environmentally friendly fuels in an economical manner. In this study, vacuum residue, a heavy fraction of crude oil, is employed as a feedstock to produce high-grade methanol. A vacuum residue to methanol (VRTM) process is simulated using Aspen Plus for a methanol production capacity of 90 t/h with 99.9 wt.% purity.

The developed VRTM process is benchmarked with the conventional steam reforming to methanol (SRTM) process through energy, environmental, and economic analysis. The performance of the vacuum residue gasifier, natural gas reformer, and methanol synthesis reactor is validated against plant data, and the simulation results are found to be in good agreement. The VRTM process offers a process energy efficiency of 49.5%, which is 1.6% higher than the SRTM process. The unit cost of methanol product from the VRTM process is \$317/tCH_3OH, which is 14% lower compared to the SRTM process.

In terms of environmental analysis, SRTM process emits less carbon emissions than the VRTM process. However, the VRTM process offers a high purity captured CO2 stream that can be utilized for another application that can further offset the methanol production cost. Moreover, the VRTM process provides a viable option to utilize vacuum residue, which is a low-value residue, as a feedstock for the production of high-grade methanol, contributing to the reduction of waste and increasing resource utilization.

Furthermore, the study assesses the potential impact of varying process parameters such as gasification temperature, pressure, and oxygen/fuel ratio on the performance of the VRTM process. The results indicate that increasing the gasification temperature and oxygen/fuel ratio can improve the process efficiency and reduce the carbon emissions. This finding highlights the potential for further optimization of the VRTM process to achieve even better environmental and economic performance.

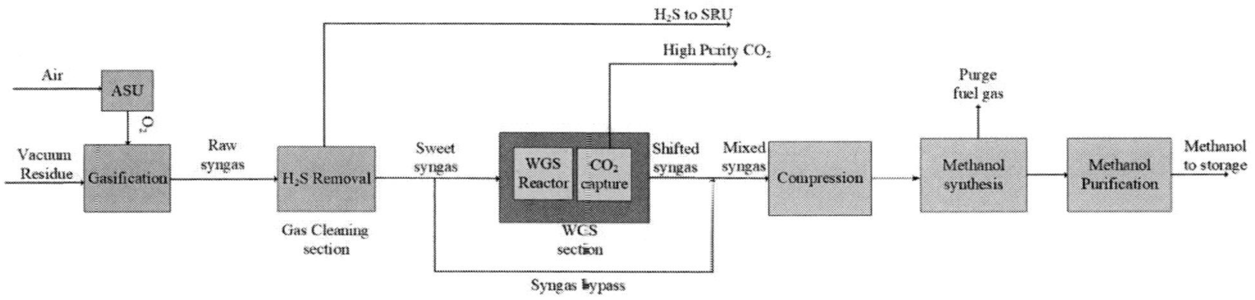

339b Decarbonization Strategies for Corn Grain Ethanol to Sustainable Aviation Fuel (SAF): Techno-Economic and Life Cycle Analyses.

Md Mosleh Uddin, Uisung Lee, Yimin Zhang, Sharon Smolinski, Arpit Bhatt, Ling Tao

Sustainable aviation fuel (SAF) grand challenge, a multiagency effort of the U.S. government, envisions producing 3 billion gallons of SAF by 2030, targeting to reach 35 billion gallons by 2050 with net-zero emissions. Achieving the near-term goal of 3 billion gallons requires utilizing existing biofuel infrastructure and adopting aggressive emission reduction strategies to meet the 50% emission reduction target in the aviation sector. SAF producers can leverage the established corn grain ethanol infrastructure in the U.S. for SAF production. The upward trend of vehicle electrification and fuel economy worldwide, including in the U.S., would shrink the ethanol blend market in the near- term leaving a significant amount of expendable ethanol for SAF upgrading.

Nevertheless, the 50% emissions reduction goals are not realizable using the existing corn grain ethanol supply chain. Current corn grain ethanol to SAF pathway can achieve only a 20% CI score reduction compared to the fossil jet fuel demanding modifications in existing corn farming practices and deployment of smart carbon reduction techniques. However, these modifications might have moderate to significant cost penalties that might be prohibitive for market penetration without policy incentives. For the successful development of large-scale fuel production, it is important to understand the economic and environmental implications of such an effort. This study integrates techno-economic analysis with life cycle assessment to study multiple deep decarbonization technology options for corn grain ethanol to SAF process concepts.

The options for reducing GHG emissions in corn grain ethanol to SAF pathway are replacing fossil NG with RNG or biomass, switching grid electricity to renewable electricity, switching to renewable hydrogen, carbon capture and storage (CCS), and carbon capture and utilization (CCU). Renewable energy sources present the potential to significantly reduce greenhouse gas (GHG) emissions (4.1–61 gCO_{2e}/MJ or 6%–86 % compared to the baseline) at the expense of significantly higher cost. Without considering transportation infrastructure-related expenses, CCS has the potential to significantly reduce GHG emissions (35 gCO_{2e}/MJ), 49% GHG reduction compared to the base case, and with only marginal increases in cost (only \$0.13/GGE increase compared to baseline).

Employing CCU is a highly capital-intensive process, and some cost penalties can be offset by a 48% increase in ethanol production. However, high electricity demand for the H_2/CO_2 electrolysis process leads to a 50% increase in MFSP. Emission reduction is only possible if renewable electricity is used for the CCU process. Given the current state of technology for CCU, further R&D efforts are necessary for commercial-scale deployment.

Since farming activities significantly impact feedstocks' life cycle GHG emissions, this study has included analysis to understand the impact of farming, soil organic carbon, and indirect land use change associated with this corn grain ethanol derived SAF. The investigated sustainable farming practices include conservative tillage, variable nitrogen rate, use of efficient fertilizer, utilizing winter cover crops, and applying livestock manure. GHG emissions reductions for sustainable farming practices vary from 22% to 34%, among which conservation tillage, variable nitrogen rate, and livestock manure have slightly lower cost than the base case. Adopting a single sustainable farming practice does not significantly reduce GHG emissions; however, the reduction contribution can be accumulated by applying multiple options to achieve considerable GHG reduction.

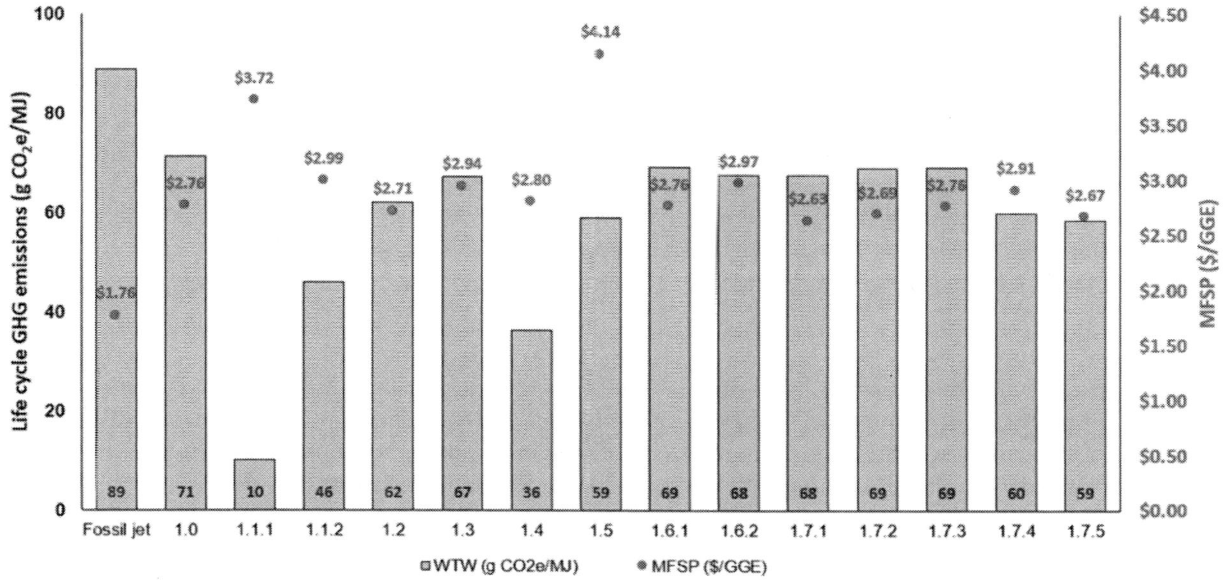

Figure 1. Summary of GHG emissions and MFSP for stand-alone corn grain ethanol to SAF baseline and all low-carbon scenarios . Case 1.0: Gen 1 baseline; 1.1.1: Process fuel switch from NG to dairy manure-derived RNG; 1.1.2: Process fuel switch from NG to biomass (forest residue); 1.2: Renewable electricity; 1.3: Green H_2; 1.4: CCS; 1.5: CCU for ethanol production with renewable electricity; 1.6.1: adding in-situ Gen 1.5; 1.6.2: adding ex-situ Gen 1.5; 1.7.1: conservation tillage; 1.7.2: variable N rate; 1.7.3: enhanced fertilizer efficiency; 1.7.4: cover crops application; 1.7.5: manure application.

339c Techno-Economic Assessment of Cerium Oxide Nanocatalyst Production Via Supercritical Hydrothermal Synthesis

Alexander Guzman-Urbina, Masazumi Tamura, Hajime Ohno, Yasuhiro Fukushima

Cerium oxide is an important material for various industrial chemical reactions due to its outstanding physicochemical properties such as high surface area, redox activity, and oxygen storage capacity[1]. Given that the nanoscale particle size of cerium oxide improves its catalytic performance due to its increased surface area, cerium oxide nano-catalysts have also gained significant attention in recent years, especially for their potential in carbon dioxide (CO_2) utilization in applications such as catalyzing the conversion of CO_2 to value-added chemicals such as polycarbonates, methanol, formic acid, etc.

Traditional cerium oxide nanocatalyst synthesis methods include chemical precipitation[2][3], spray pyrolysis[4], and solvothermal method[5]. These methods usually require long reaction times, and the use of hazardous chemicals, resulting in high energy demand and environmental impacts[6]. Supercritical hydrothermal synthesis (SHS) is an emerging method for the sustainable synthesis of cerium oxide nanocatalyst. This method involves the reaction between a cerium salt and water under supercritical conditions, leading to enhanced mass transfer and reaction rates, resulting in the formation of high-quality nanoparticles[7] (Fig.1). Despite the promising results of the supercritical hydrothermal synthesis, there are still challenges that need to be addressed. One of the primary challenges is the scalability of the process. While supercritical hydrothermal synthesis has been demonstrated on a laboratory scale, it is essential to evaluate its feasibility for large-scale production. The techno-economic assessment of the industrial-scale production of cerium oxide nanocatalyst using supercritical hydrothermal synthesis presented in this study addresses this challenge. The results demonstrate the potential of the process for scalability with acceptable values, paving the way for further research and development in this area.

The study was divided into two main parts. First, a preliminary evaluation was conducted to determine the availability and production of the cerium oxide mineral. Second, a mass and energy balance, operating costs, and capital investments were estimated for the scalability of the supercritical hydrothermal synthesis process. The process was divided into four main subsystems: metal salt solution, nanoparticles formation, separation and solvent recovery, and auxiliary subsystems for heating and cooling utilities.

Preliminary results showed that for an industrial production of 42 ton/day of cerium oxide nanocatalyst, the heating consumption would be 3.4 MW/ton, and the minimum selling price would be 3.2 kUSD/ton. Furthermore, a sensitivity analysis was performed on the precursors' concentration to assess their impact on energy consumption. The results demonstrated the potential of supercritical hydrothermal synthesis for scalability with acceptable values in comparison to current market trend of nanomaterials. In conclusion, the sustainable synthesis of cerium oxide nanocatalysts is crucial for the development of environmentally friendly and economically viable industrial chemical reactions, including CO_2 utilization. The findings of this study provide insights into the feasibility of supercritical hydrothermal synthesis for the large-scale production of cerium oxide nanocatalyst, highlighting its potential as a sustainable and cost-effective alternative to traditional synthesis methods.

[1] Zhang, Q., et al. 'Recent advances in cerium oxide-based nanomaterials for catalysis and energy applications.' Chemical Society Reviews 45.11 (2016): 3177-3216.

[2] Zaboeva, E. A., Izotova, S. G., & Popkov, V. I. (2016). Glycine-nitrate combustion synthesis of CeFeO 3-based nanocrystalline powders. *Russian Journal of Applied Chemistry*, *89*, 1228-1236

[3] Keerthana, M., Malini, T. P., & Sangavi, R. (2022). Efficiency of cerium oxide (CeO2) nano-catalyst in degrading the toxic and persistent 4-nitrophenol in aqueous solution. *Materials Today: Proceedings*, *50*, 375-379

[4] Chaisuk, C., Wehatoranawee, A., Preampiyawat, S., Netiphat, S., Shotipruk, A., &

Mekasuwandumrong, O. (2011). Preparation and characterization of CeO2/TiO2 nanoparticles by flame spray pyrolysis. *Ceramics International, 37*(5), 1459-1463

[5] Sun, C., Li, H., Zhang, H., Wang, Z., & Chen, L. (2005). Controlled synthesis of CeO2 nanorods by a solvothermal method. *Nanotechnology, 16*(9), 1454

[6] Damanik, M., Febriyanti, F., Hofifah, S. N. & Nandiyanto, A. Synthesis of CeO2 Nanoparticles by Various Methods. Arabian Journal of Chemical and Environmental Research 08, 382–417 (2021).

[7] Adschiri, T., Kanazawa, K., & Arai, K. (1992). Rapid and continuous hydrothermal crystallization of metal oxide particles in supercritical water. Journal of the American Ceramic Society, 75(4), 1019-1022.

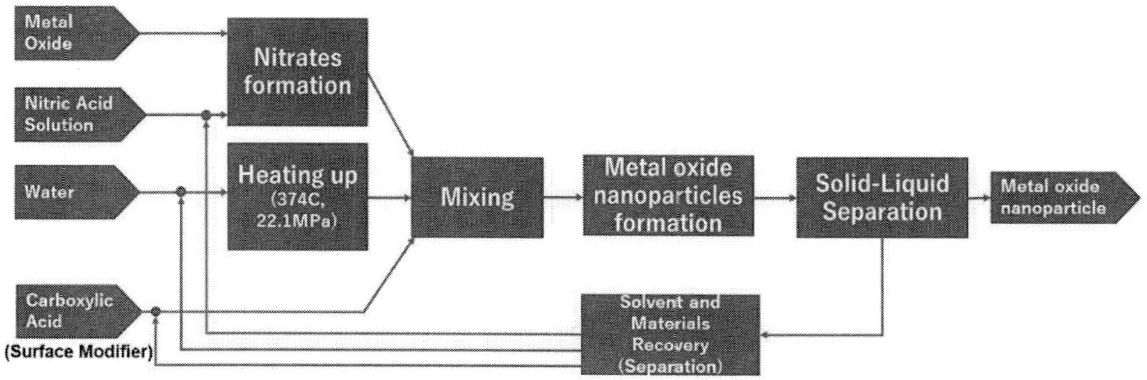

Figure 1. Simplified block diagram for cerium nanocatalyst production by the SHS method

339d Ammonia As an Alternative Energy Carrier for Electricity Generation in India

Lingyan Deng, Guiyan Zang, Angiras Menon, Dharik Mallapragada, Emre GenÃ§er, Ahmed Ghoniem, William Green, Robert J. Stoner

The advantages of higher hydrogen density, easier liquification, and capable of co-combust with coal makes ammonia an attractive energy storage and carrier chemical for power generation domestically and overseas. India's coal contributes to 81% of electricity generation as of 2021. In this work, two decarbonization method are been compared economically and environmentally. One is the ammonia co-firing with coal, the other is using carbon capture and storage (CCS). Ammonia is supposed to be produced from Australia and then transported to India to co-firing with coal in power plants. One blue and one green ammonia production pathways are been considered. Aspen Plus models were built for the power plants for mass and energy balance. The cost analysis was based on greenfield plants. The results show that the levelized cost of electricity and greenhouse gas emissions of 20% blue ammonia co-firing with coal was estimated to be \$90.5/MWh and 1191 kg CO_2 e/MWh, respectively. While the green ammonia co-firing with coal was estimated to be \$150/MWh and 1597 kgCO_2e/MWh, respectively. Green ammonia is more expensive and emits more CO_2 mainly because the grid electricity price and carbon intensity are high. When using wind power, where electricity price drop to \$40/MWh, the blue ammonia co-firing case would results in \$88.6/MWh and 1079 kg CO_2e/MWh; while the green ammonia co-firing case would results in \$124.6/MWh and 1062 kg CO_2e/MWh. Coal-firing with 20% carbon capture would results in \$68/MWh and 1226.6 kgCO_2e/MWh. In general, if technology and location available, CCS is less expensive than ammonia co-firing for decarbonization. However, if we also consider the geological availability of CO_2 storage, government regulations, social acceptance of CO_2 storage, there are additional constraint for the CCS decarbonization method. In addition, the technology readiness level of CCS in India is relatively low compared to the U.S. or European countries, and there is high energy penalty, the net plant efficiency would drop by about 8% for the CCS case, while ammonia co-firing case only has slight efficiency drop (~1%). Another advantage is that the ammonia co-firing case requires relatively less modification for the plant, and it could reduce the direct emission. No additional construction is needed for downstream CO_2 transportation and storage. As an chemical energy storage carrier, ammonia could further help provide a sustain electricity grid.

339e Techno-Economic Analysis and Life Cycle Assessment for Production of Biodegradable Polyesters

Bo-Xun Wang, Brian Grady, George Huber, Victor Zavala

Plastics in the ocean water pose a significant threat to marine animals, causing entanglement or being eaten mistakenly. Most plastic waste enters rivers and the ocean because of the inappropriate disposal. In 2010, 8 million tonnes of plastic leaked into the ocean, which accounts for 3% of global annual plastic waste [1]. One possibility to reduce or eliminate this problem is to use biodegradable plastics (BPs). Biodegradable plastics can be decomposed to innocuous substances in the natural environments through chemical hydrolysis or enzymatic action [2]. In this way, the plastic marine pollution problem can be mitigated as more plastics are replaced by BPs.

As people are more aware of plastic marine pollution, the global consumption of BPs has increased rapidly since their first introduction in 1980s [3]. BPs can be applied in many different fields such as medicine, packaging, and agriculture [4–6]. Among various types of BPs, aliphatic polyesters are studied extensively due to their highly hydrolysable ester bonds. They are synthesized through polycondensation between diols and dicarboxylic acids under high vacuum and high temperature condition. However, aliphatic polyesters suffer from poor mechanical and thermal properties compared to fossil-fuel based LDPE and LLDPE [7]. To tackle this problem, an aromatic diacid is selected and added as co-monomer into the process. The aromatic parts in polyester can enhance mechanical and thermal properties, but also decrease the biodegradability of polymer [8]. Among aliphatic-aromatic copolyesters, poly(butylene adipate-co-terephthalate) (PBAT) is currently one of the most promising polymers for applications due to its good flexibility and biodegradability [9]. PBAT has been applied in a wide range of packaging applications, and PBAT is commercially known as Ecoflexâ and was first prepared by BASF in the 1990s.

In this work, we propose new polyesters that are promising to replace the conventional biodegradable polyester, PBAT. Since to our knowledge there is no comprehensive technoeconomic analysis (TEA) or life-cycle assessment (LCA) on the production of PBAT in the open literature, TEA and LCA for the targeted polyesters have been conducted as well. The minimum selling price (MSP) for the new polyester containing furan-dicarboxylic acid and 1,5-pentanediol as monomers is determined to be \$2.02/kg, which is 20% lower than that for PBAT (\$2.52/kg). The sensitivity analysis of MSPs for the polyesters are performed by changing the various factors that can impact process economy, including production scale, feedstock prices, and financial parameters. In addition, we conduct the LCA on the production of the targeted polyesters, selecting global warming, energy consumption, and fossil resource depletion as representative indicators. Our analysis shows that the biomass-derived polyesters can have about 50% lower of carbon footprint and fossil resource depletion compared to the petroleum-derived polyesters. The production for 1 kg of biomass-derived poly(pentylene adipate-co-furandicarboxylate) will cause 3.05 kg CO_2-equivalent of GHG emission.

References:

[1] H. Ritchie, M. Roser, Plastic pollution, Our World in Data. (2018).

[2] R.A. Gross, B. Kalra, Biodegradable Polymers for the Environment, Science (1979). 297 (2002) 803–807.

[3] I. Vroman, L. Tighzert, Biodegradable Polymers, Materials 2009, Vol. 2, Pages 307-344. 2 (2009)

307–344.

[4] Y. Ikada, H. Tsuji, Biodegradable polyesters for medical and ecological applications, Macromol. Rapid Commun. 21 (2000) 117–132.

[5] M.I. Din, T. Ghaffar, J. Najeeb, Z. Hussain, R. Khalid, H. Zahid, Potential perspectives of biodegradable plastics for food packaging application-review of properties and recent developments. 37 (2020) 665–680.

[6] M.A.M. Akhir, M. Mustapha, Formulation of Biodegradable Plastic Mulch Film for Agriculture Crop Protection: A Review, 62 (2022) 890–918.

[7] T. Kaneko, T.H. Thi, D.J. Shi, M. Akashi, Environmentally degradable, high-performance thermoplastics from phenolic phytomonomers, Nature Materials 2006 5:12. 5 (2006) 966–970.

[8] R.J. Müller, I. Kleeberg, W.D. Deckwer, Biodegradation of polyesters containing aromatic constituents, J Biotechnol. 86 (2001) 87–95.

[9] J. Jian, Z. Xiangbin, H. Xianbo, An overview on synthesis, properties and applications of poly(butylene-adipate-co-terephthalate)–PBAT, Advanced Industrial and Engineering Polymer Research. 3 (2020) 19–26.

339f Design and Scale-up of a Multiple Production Process of Food, Cosmetics and Biofuel from Unused Oil Derived from Rice Bran

Naomi Shibasaki-Kitakawa, Kousuke Hiromori, Makiko Kato

Many technologies have been developed to utilize the large amounts of unused bioresources generated by existing industries, but only a few have been put into practical use. This is due to the difficulty of achieving both environmental compatibility and economic efficiency. We believe this is because the way the current manufacturing process is designed is not suitable for utilizing unused bioresources. So far, the focus has been on innovative products and designing processes to manufacture them efficiently. In this case, a process is designed to first select suitable raw materials for production, extract the desired components, efficiently convert them and remove impurities to make the product. However, the separation and purification process to remove unwanted components is a barrier to commercialization due to its high environmental impact and cost.

We have changed our thinking, focusing on components contained in unused bioresources and proposing a new method of designing processes that do not generate unwanted substances. The aim of this method is to evaluate the market value of each component, recover the higher ones in high purity and convert the lower ones into higher value-added products, thereby producing as much product as possible and improving the overall economics of the process.

In this research, we have applied this concept to unused oil generated in large quantities at rice bran oil refinery, and have successfully developed a practical device for a multiple production process to recover several functional ingredients at high purity, and to convert the remainder into biofuels. In traditional industries, these products are manufactured in separate processes from different raw materials. In contrast, our process allowed the production of all products from one unused bioresource without generating any unwanted components.

Our process utilizes three ion exchange resins as new catalysts and adsorbents in an oil solution system, allowing complete conversion of fatty acids and triglycerides to fatty acid esters without constraints of reverse reactions at 50°C and atmospheric pressure[1)]. At the same time, by-products of water and glycerol are physically adsorbed onto the resin and removed from the solution, and phenolic compounds such as vitamin E are selectively retained in the resin by ion exchange[2)].

Furthermore, the solution that flowed out of the resin packed reactor spontaneously formed a white precipitate when cooled to room temperature, which was collected by filtration and found to be paraffin with a purity of over 95% and a card structure, used as base materials for lipstick. This is the world's first biomass-derived paraffin, which has acquired its own CAS number as well as an international patent[3)]. Further distillation of the remaining alcohol precipitated phytosterols, which could be recovered with a purity of over 90%, and the remaining liquid was directly used as biofuel for power generation.

This process has already been commercialized by a Tohoku University start-up[4)], and the products are sold both domestically and internationally. This technology can also be applied to deodorized distillates of various other vegetable oils, and can contribute significantly to the creation of a resource cycle for unused oils.

1)N.Siddique *et al., Fuel*, 289, 119884 (2021)

2)K.Hiromori *et al., J.Clean. Prod.*,189,223(2018)

3)K.Hiromori *et al.,* WO2023/021593(2023)

4)https://phytochem-products.co.jp/en/

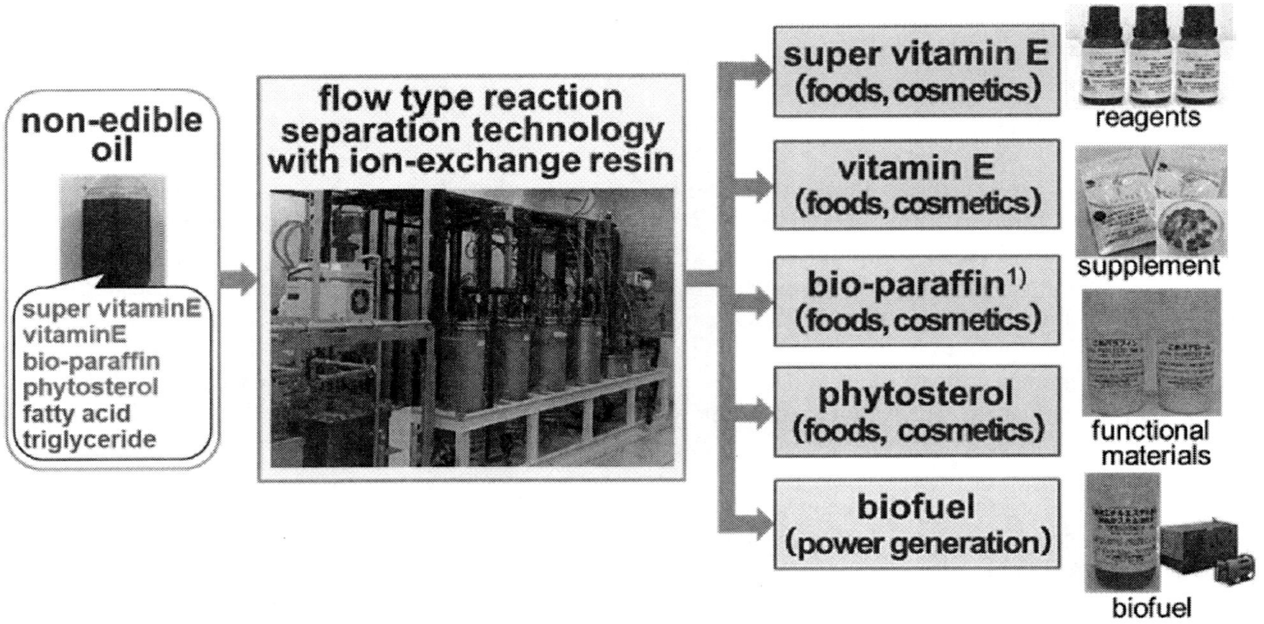
non-edible oil
super vitaminE
vitaminE
bio-paraffin
phytosterol
fatty acid
triglyceride
flow type reaction separation technology with ion-exchange resin
super vitamin E (foods, cosmetics)
vitamin E (foods, cosmetics)
bio-paraffin1) (foods, cosmetics)
phytosterol (foods, cosmetics)
biofuel (power generation)
reagents
supplement
functional materials
biofuel

339g Techno-Economic Analysis of Antisolvent Crystallization for Inorganic Salt Recovery

Barnabas Gao, Parul Sahu, Kirti Yenkie, Gerard Capellades

Introduction

Natural brines and other mineral sources contain multiple inorganic salt components such as lithium (Gao et al., 2012), iron and base-metal-rich sulfide, magnesium, and chloride (Nishri et al., 1988). However, recovery of these inorganic salts from aqueous saline systems is a challenging task due to their high solubility (Foo et al., 2022). Technologies for recovering these salts from the concentrated brines are limited and in most cases, the concentrated brine solutions are ejected into the sea (Vallès et al., 2023). Precipitation and crystallization-based techniques have been exploited commercially for inorganic salt recovery from natural brines and mineral reserves. Recent studies in the crystallization space indicate antisolvent crystallization as a promising technique that offers several advantages over traditional methods (Moldoveanu and Demopoulos, 2015). Some of the advantages of antisolvent crystallization include, low energy requirement, ambient operating temperature, and the capability to recover and reuse solvents rather than discharging back to the ecosystem, thereby leading to reduced costs and environmental impacts (Raj R and kurup, 2016). In this study, we analyzed the techno-economics of an antisolvent crystallization process for the recovery of sodium chloride and sodium sulphate from bittern.

Methodology

During salt production via desalination process, a major byproduct is the concentrated aqueous solution of multiple salts, also known as bittern. This bittern is rich is sodium sulphate as well as leftover sodium chloride (Battaglia et al., 2023).

To understand the recovery of these salts selectively from bittern we first conducted laboratory experiments to determine the process conditions for optimal crystallization yield. We used several candidate antisolvents such as ethanol, methanol, acetone, acetonitrile, ethylene glycol and dimethyl sulfoxide and based on the dissolution test runs, we evaluated and chose ethanol for the study. Ethanol is chosen as the solvent due to its potential for high crystallization yields and its relatively low environmental impact as a greener organic solvent.

We further conducted a detailed mass balance, and energy consumption estimates for the process. We estimated the total product cost which encapsulated the raw material cost, utility cost, overhead cost (other cost), labor cost, equipment cost and cost of consumables. The cost of the equipment and other consumables were taken from SuperPro Designer databanks using the base cost analysis for the year 2023. The minimum weighted selling price for the recovered salts (NaCl and Na_2SO_4) required to achieve profit is estimated and thus the break-even profitability metric was used in the analysis. The process schematic is represented in Figure 1.

Results

The economics and process sustainability of the antisolvent crystallization process for selective inorganic salt recovery from concentrated aqueous saline is presented in this work. The economic viability of the inorganic salt (NaCl and Na_2SO_4) recovery process with a capacity of 8 tons/day is evaluated with an annualized capital cost evaluated using a capital recovery factor of 0.11, assuming a plant life of 25 years.

The evaluated cost distribution for the process shown that the total cost for operation is highly influenced by the utility cost and the cost of the solvent. The total production cost is analyzed for the process with solvent recovery and also with no solvent recovery.

The raw material cost for the recovery process is calculated from the equation

Raw material (RM) cost = cost of RM – cost of RM (recovery) = $C_I{*}M_J$ - $C_I{*}rM_J$

Overall, this study contributes to the development and the adoption of circular economy principles, which is an economic model that aims to reduce waste and promote the efficient use of resources. The circular economy model is based on the principles of reducing, reusing, and recycling materials to create a closed loop system. Antisolvent crystallization with solvent recovery contributes to circular economy by enabling the reuse of solvents, reducing waste generation and conserving resources. Additionally, the recovered solvent is reused in subsequent processes, reducing the need for new materials and promoting a closed loop system.

REFERENCES

Battaglia, G., Domina, M.A., Lo Brutto, R., Lopez Rodriguez, J., Fernandez de Labastida, M., Cortina, J.L., Pettignano, A., Cipollina, A., Tamburini, A., Micale, G., 2023. Evaluation of the Purity of Magnesium Hydroxide Recovered from Saltwork Bitterns. Water 15, 29. https://doi.org/10.3390/w15010029

Foo, Z.H., Stetson, C., Dach, E., Deshmukh, A., Lee, H., Menon, A.K., Prasher, R., Yip, N.Y., Lienhard, J.H., Wilson, A.D., 2022. Solvent-driven aqueous separations for hypersaline brine concentration and resource recovery. Trends Chem. 4, 1078–1093. https://doi.org/10.1016/j.trechm.2022.09.004

Gao, F., Zheng, M., Song, P., Bu, L., Wang, Y., 2012. The 273.15-K-Isothermal Evaporation Experiment of Lithium Brine from the Zhabei Salt Lake, Tibet, and its Geochemical Significance. Aquat. Geochem. 18, 343–356. https://doi.org/10.1007/s10498-012-9168-1

Moldoveanu, G.A., Demopoulos, G.P., 2015. Organic solvent-assisted crystallization of inorganic salts from acidic media: Crystallization of inorganic salts with 2-propanol. J. Chem. Technol. Biotechnol. 90, 686–692. https://doi.org/10.1002/jctb.4355

Nishri, A., Herbert, H.J., Jockwer, N., Stichler, W., 1988. The geochemistry of brines and minerals from the Asse Salt Mine, Germany. Appl. Geochem. 3, 317–332. https://doi.org/10.1016/0883-2927(88)90109-6

Raj R, A., kurup, M., 2016. ANTISOLVENT CRYSTALLIZATION: A NOVEL APPROACH TO BIOAVAILABILITY ENHANCEMENT. Eur. J. Biomed. Pharm. Sci. 3, 230–234.

Vallès, V., López, J., Labastida, M.F. de, Gibert, O., Leskinen, A., T. Koivula, R., L. Cortina, J., 2023. Polymeric and inorganic sorbents as a green option to recover critical raw materials at trace levels from sea saltwork bitterns. Green Chem. 25, 700–719. https://doi.org/10.1039/D2GC02338E

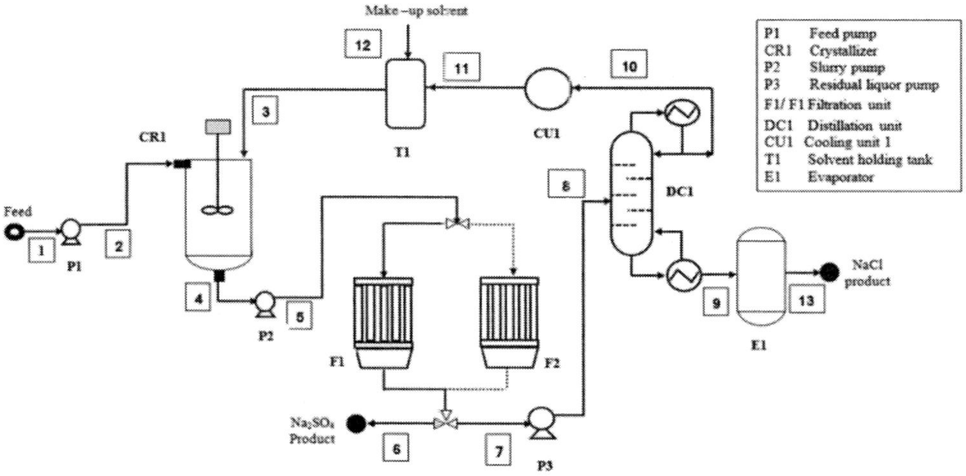

Figure 1:Process flow diagram of the antisolvent crystallization process (CR1), including two parallel batch filters (F1/F2) and evaporator (E1) for salt recovery, and a distillation step (DC1) for solvent recovery.

339h Development of an Autothermal Integrated Methane Pyrolysis with Chemical Looping Combustion for Onsite Hydrogen Production

Jude Okolie

The limitations of hydrogen pipeline infrastructure especially between production plants and fueling as well as the issues of hydrogen embrittlement have stimulated interest in alternative hydrogen production pathways. A mobile autothermal methane pyrolysis unit is a promising pathway for hydrogen production that could help mitigate the challenges of hydrogen transportation and storage. The present study presents a novel methane pyrolysis integrated with chemical looping combustion for efficient hydrogen production. The process was simulated with Aspen plus and detailed techno-economic and life cycle assessment (LCA) were performed to evaluate the process economics and environmental impact. The introduction of chemical looping combustion increased the energy efficiency up to 92%. Additionally, the levelized cost of hydrogen for the proposed technology ranges between \$1.3 U.S.D/KG - \$1.5 U.S.D/kg. A cradle to gate LCA analysis was performed and compared with a standalone methane pyrolysis unit. LCA results shows that the process is environmentally benign compared to a stand-alone methane pyrolysis system. Additionally, a stochastic sensitivity analysis indicates that factors such as tax rate, catalyst and equipment purchase cost influenced the levelized cost of hydrogen.

339i Economic Analysis of a Power Plant Under Health Monitoring in an Elastic Energy Market

Chandra Sekhar Somayajula, Debangsu Bhattacharyya, Xingbo Liu, Shanshan Hu

Renewables are projected to be the primary source of electricity generation in the future. But fossil energy sources like coal, natural gas, etc., will still be prevalent in the foreseeable future and will be in use to ensure the stability an resiliency of the electricity grid by assisting in meeting the load. Due to this, coal-fired power plants (CFPP) are being subjected to load-following because of the increasing penetration of intermitted renewables. The existing supercritical pulverized coal-fired power plants are generally designed to operate at base-loaded conditions.[1] From 2012 to 2017, on average, about 6.4% of annual potential production lost due to forced outages was caused by waterwall failure, making it the top cause for loss of revenue. Several mitigation steps have been recommended for CFPP to minimize damage and operation and maintenance costs. One approach is to monitor the health of the critical components by using a network of sensors. Hot corrosion of critical boiler components is one of the key reasons for the tube failures and therefore this work particularly focuses on monitoring of hot corrosion in high temperature locations. Measuring hot corrosion in the challenging boiler environment, but recently a novel electrochemical sensor has been developed in-house for in-situ corrosion monitoring.[2] However, these sensors are novel and costly and therefore a return on investment be justified. The improved revenue due to the anticipated reduction in forced outages due to corrosion monitoring must justify the purchase, installation, and maintenance costs for sensors. However, the improvement in the revenue depends on several factors. First of all, the forced outage adds to the maintenance cost and reduces the capacity factor, which can increase the production cost for unit energy. The increase in availability due to a reduction in forced outage needs to be computed with due consideration of future energy market instead of historical data, i.e., elasticity in the energy market. Due to the anticipated ongoing increase of renewable power in the electric grid, they are expected to pick up the lost power production from the plant that went offline due to a forced outage. Therefore, improvement in availability may not proportionately lead to an increase in the number of operating hours or increased generation. Furthermore, measurement of variables such as temperature, and composition of the flue gas can provide indirect estimate of the rate of corrosion and therefore may be used in conjunction with the corrosion sensors thus enabling use of fewer corrosion sensors in the boiler. A scenario-based economic analysis and payback period analysis are conducted in this work to estimate the optimal investment in the sensor network.

In the existing literature, there is work on economic analysis of traditional CFPP using measures like net present value (NPV).[3] Economic analysis of CFPP combined with other technologies like solar energy, biomass gasification, CO_2 capture etc., have been conducted.[4]–[6] To the best of our knowledge, there is no work in the literature where the impact of the increased availability of CFPP on its net power generation has been analyzed by accounting for market elasticity.

For analyzing future of market, an energy market forecasting software "TIMES" is used.[7] TIMES provides the capability of analyzing various scenarios such as the impact of reduction in O&M cost availability etc., of coal-fired power plants on the increase in their power production. Changes in energy trends are incorporated into the model through scenario files. Technologies and corresponding factors significantly affecting electricity production of coal-fired power plants are identified.[8] Impacts of each of these technologies on the electricity production from the coal-fired power plant are analyzed. Using Latin Hypercube Sampling (LHS), the set of near-random scenarios that include all electric energy generation technologies are created. The electricity produced by coal-fired power plants with and without improved availability under these random scenarios is calculated and analyzed. Potential revenue gained by plants under increased availability is estimated. In-house data and information available in the open literature are used to calculate the investment cost for the sensor network. Finally, a comparison is made between the sensor network cost and potential revenue gain to determine the economic feasibility.

References

[1] G. Grol, Eric; Tarka, Thomas J.; Myles, Paul; Bartone, Jr, Leonard M.; Simpson, James; Rossi, "Impact of Load Following on the Economics of Existing Coal-Fired Power Plant Operations," 2015.

[2] N. N. Aung and X. Liu, "High temperature electrochemical sensor for in situ monitoring of hot corrosion," *Corros. Sci.*, vol. 65, pp. 1–4, 2012, doi: 10.1016/j.corsci.2012.08.010.

[3] R. Kumar, A. K. Sharma, and P. C. Tewari, "Thermal performance and economic analysis of 210 MWe coal-fired power plant," *J. Thermodyn.*, vol. 1, no. 1, 2014, doi: 10.1155/2014/520183.

[4] C. Li, R. Zhai, B. Zhang, and W. Chen, "Thermodynamic performance of a novel solar tower aided coal-fired power system," *Appl. Therm. Eng.*, vol. 171, no. November 2019, p. 115127, 2020, doi: 10.1016/j.applthermaleng.2020.115127.

[5] B. Huang *et al.*, "Industrial test and techno-economic analysis of CO2 capture in Huaneng Beijing coal-fired power station," *Appl. Energy*, vol. 87, no. 11, pp. 3347–3354, 2010, doi: 10.1016/j.apenergy.2010.03.007.

[6] H. Chen *et al.*, "Thermo-Economic analysis of a novel biomass Gasification-Based power system integrated with a supercritical CO2 cycle and a Coal-Fired power plant," *Energy Convers. Manag.*, vol. 266, no. March, pp. 1–18, 2022, doi: 10.1016/j.enconman.2022.115860.

[7] A. Mirakyan and R. De Guio, "Integrated energy planning in cities and territories: A review of methods and tools," *Renew. Sustain. Energy Rev.*, vol. 22, pp. 289–297, 2013, doi: 10.1016/j.rser.2013.01.033.

[8] J. E. T. Bistline and D. T. Young, "Economic drivers of wind and solar penetration in the US," *Environ. Res. Lett.*, vol. 14, no. 12, p. 124001, 2019, doi: 10.1088/1748-9326/ab4e2d.

339j Rotating Packed Bed (RPB) for Modular CO_2 Capture: Design and Operation Optimization

Howoun Jung, Jay Hyung Lee

Rotating packed bed (RPB) is an intensified unit that enhances mass transfer and unit throughput with rotation of packing inside the column. The rotation enables enlarged contact areas between liquid and gas phases and use of high-viscosity solvents, affording a significant reduction in the size and capital cost of separation units. A compact and modularized CO_2 capture process is an exemplary application of RPB. Several experiments and simulation studies support the advantages of applying RPB to the CO_2 capture process, mainly the size reduction of large absorption and desorption columns in a conventional process. The challenge is that there currently is no reported case of commercial-scale application of RPB to the CO_2 capture process, and therefore, it is difficult to examine its viability. In particular, the energy efficiency of the CO_2 capture process can be significantly affected by the design of the RPB units (e.g. inner and outer radii, height) as well as the operating conditions since the mass transfer area and hydrodynamics inside the column vary along the radial axis. Therefore, to fully examine its potential, efficient RPB unit design and operation conditions should first be explored with rigorous modeling and simulation.

This study addresses a simultaneous optimization of the design and operation condition of the CO_2 capture process using RPB columns. A process model for RPB columns with 30wt% Monoethanolamine (MEA) is first constructed in commercial software of gPROMS. Then, the optimal RPB design and operation of the capture process for industrial-scale CO_2 emission sources are explored through optimization, which minimizes the annualized CO_2 capture cost with prudently chosen constraints (e.g. hydrodynamics). By simultaneously optimizing the design and operation condition variables as decision variables, we explore the optimal unit design and operation of a modular CO_2 capture process with RPB.

388a Efficient Removal of Micro- and Nano-Plastics from Water Using Ionic Liquid-Based Liquid-Liquid Extraction

Ashish Srivastava, Jason Bara, Milad Esfahani

In the 21st century, Micro- and Nano-plastics (MNPs) are the potential emerging plastic-related pollutants that have adverse impacts spreading across all the environmental media, including fresh water. The current conventional water treatment methods, such as coagulation, flocculation, filtration, photocatalytic degradation, and microbial bioremediation, showed a lack of efficiency in the removal of different MNPs. Hence, new removal techniques are needed to be developed. The Liquid Liquid extraction (LLE) method is a promising separation technique for efficiently removing MNPs in water treatment. In the present work, we report the removal of two types of polystyrene (PS) MNPs from water, including synthesized micron-size (>1 μm) PS and commercial sulfonated PS (nano-size, ~70 nm) using a hydrophobic IL [C_8mim][Tf_2N]. The extraction process showed more than 99% removal of both PS MNPs at the ratio of 1:40 (vol:vol) IL/water for the removal of 100 ppm PS MNPs from water. The extraction process using other volume ratios revealed 98.89%, 96.26%, and 98.60 % removal for the ratios of 1:100, 1:200, and 1:1000 (vol:vol) IL/water, respectively. The kinetic data showed that the maximum adsorption reached after 3 h. In addition, the extraction efficiency of PS MNPS was evaluated in the presence of 2000 ppm NaCl, and ~ 99.36 % removal was achieved, showing the selectivity of [C_8mim][Tf_2N] towards PS MNPs

388b Phase Equilibria and Diffusivities of HFC-32 and HFC-125 in Encapsulated Ionic Liquids

Kalin R. Baca, Abby Harders, Nicholas Starvaggi, Yifei Wang, Lucia Matamoros Valenciano, Madelyn Bennett, Dorothy M. Haggard, Irene Xu, Emily Pentzer, Mark B. Shiflett

Todays generation of refrigerants, hydrofluorocarbons (HFCs), are being phased out due to high global warming potential (GWP) by the American Innovation and Manufacturing Act. HFCs are complex refrigerant mixtures that are unable to be separated by conventional technology because majority of the mixtures form azeotropes. With 850 million kilograms of refrigerant used in cooling applications, millions of kilograms of HFC mixtures will need to utilize new technology to be efficiently recycled and repurposed at their end of life. One separation solution being investigated is the use of ionic liquids (ILs) as entrainers in extractive distillation to selectively separate azeotropic HFC mixtures. ILs show promise for the separation due to their unique properties including negligible vapor pressure, thermal and chemical stability, and solubility of many gases. Recently, IL/polymer combinations such as mixed matrix membranes, supported ILs, and encapsulated ILs (ENILs) have been investigated for a variety of applications. ENILs in particular are a new technology showing promise as a stable, re-usable, and static solution that is still able to take advantage of the IL properties while also having an increased mass transfer rate. This presentation will discuss the first solubility measurements of HFC-32 (difluoromethane, CH_2F_2) and HFC-125 (pentafluoroethane, C_2HF_5) in encapsulated ILs (ENILs) to investigate their potential as a solution for separating azeotropic refrigerant mixtures. The ENIL results are compared to solubility measurements in neat ILs. Phase equilibria data for HFC-32 and HFC-125 in both the ENILs and neat ILs were measured using a Hiden Isochema IGA gravimetric balance.

388c Pilot-Scale Extractive Distillation System to Separate the Azeotropic Mixtures of HFC-32 and HFC-125, from Recycled Refrigerant, R-410A Using Ionic Liquid [C_1C_2im] [Tf_2n]

Julia Espinoza Mejia, Mark B. Shiflett, Aaron Scurto

Objectives/Scope

Hydrofluorocarbons (HFCs) are potent greenhouse gases that accelerate anthropogenic global climate change. Though the U.S. Environmental Protection Agency (EPA) proposed a rule under the American Innovation and Manufacturing (AIM) Act of 2020 to significantly reduce the production and consumption of these climate-damaging pollutants, today, an estimated 2,800 kilotons global refrigerant supply will need to be effectively and responsibly disposed of. This investigation presents the optimization of a pioneer, cutting-edge, in-house-built pilot-scale extractive distillation system fully instrumented and automatized to separate the azeotropic mixture of R-410A. This technology uses recycled refrigerant, R-410A (50/50 wt% of HFC-32 and HFC-125). The HFC-32/HFC-125 azeotropic blend is separated using 1-Ethyl-3-methylimidazolium bis (trifluoromethyl sulfonyl) imide [C_1C_2im][Tf_2N] as an ionic liquid entrainer.

Methods/Procedures/Process

A state-of-the-art extractive distillation system will be discussed that comprises a tower with structured packing, reboiler, flash tank, condenser, heat exchanger, reflux drum, pumps, and other minor equipment and instrumentation. Likewise, a gas chromatograph with a thermal conductivity detector and helium carrier gas is used to verify the tested refrigerant products' compositions. Initially, 3 kg/h of refrigerant R-410A (50/50 wt% of HFC-32 and HFC-125) is fed into the column as either a vapor or liquid at 20 °C. The column operates at 0.9 MPa, and its stripping and rectifying sections are split into four symmetric structured packings with 12 embedded RTDs. A camera and LED light are placed in each sight glass of the column to observe the liquid distribution entering the packing. Each packing has pressure drop sensors to determine the column hydraulics and flooding regions. Above the column packing sections, a single pass shell-tube vertical condenser collects the condensate into a reflux drum.

The facility chilling water is the cooling fluid that exchanges heat with the light product HFC-125 in the condenser. HFC-125 production rate is fixed, and the rest of the accumulated condensate is returned to the column; the proportionality of the flow meters calculates the reflux ratio. In the bottom part of the column, a partial reboiler is installed where the solvent [C_1C_2im][Tf_2N] and the heavy product HFC-32 leaves to feed a flash tank for ionic liquid recovery. A proportional valve or a diaphragm pump, aligned to the outlet stream, controls the reboiler liquid level. The flash tank operates at vacuum conditions of 0.01 MPa and 100 °C drawn by a vacuum pump. A demister is attached to the flash tank vapor stream to prevent solvent in the recovered HFC-32. In the flash tank, the liquid level is controlled by the outlet liquid flow through a diaphragm pump back to the column. A cooling water heat exchanger is used before the solvent returns to the column. Since the light product obtained from the flash tank condenses at low temperatures, a dry-ice bath is used before collecting the product in a refrigerant cylinder. All digital readings are recorded with the National Instruments LabVIEW software. All the vessels and internals are constructed with 304/304L stainless steel, which offers the most significant resistance to ionic liquid corrosion.

Results/Observations/Conclusion

Heavy and light refrigerant products are obtained from the extractive distillation system presented in this investigation work. The light product HFC-125 is produced with a purity of 99.5 wt%; the rest is HFC-32 in 0.5 wt%. On the other hand, the heavy product HFC-32 is recovered with 95.5 wt% purity, and the rest corresponds to HFC-125 in 0.5 wt%. The HFC-125 recovered refrigerant meets the technical specifications (AHRI 700) for the industrial commercialization and future reuse as a fluorinated feedstock for producing new low global warming potential products such as fluoropolymers, agriculture chemicals, and electrolytes for batteries. Similarly, the produced HFC-32 recovered refrigerant meets the technical specifications and can be repurposed into future refrigerant products such as R-454B (i.e., 68.9 wt.% HFC-32 and 31.1 wt% 2,3,3,3-tetrafluoropropene, HFO-1234yf) due to the low global warming potential.

Application/Significance/Novelty

This experimental work is the first attempt to separate HFC-32 and HFC-125 refrigerants by using the ionic liquid [C_1C_2im] [Tf_2N] in a pilot-scale extractive distillation system. Likewise, the technology presented corresponds to the first pressurized extractive distillation tower designed for using ionic liquid [C_1C_2im] [Tf_2N] as an entrainer. This innovation includes a continuous ionic liquid solvent recovery system with a vacuumed flash vessel. The results from this effort provide significant support to the industrial application of azeotropic separation with ionic liquid entrainer in extractive distillation for recycling HFC refrigerants.

Figure 1

Process flow diagram

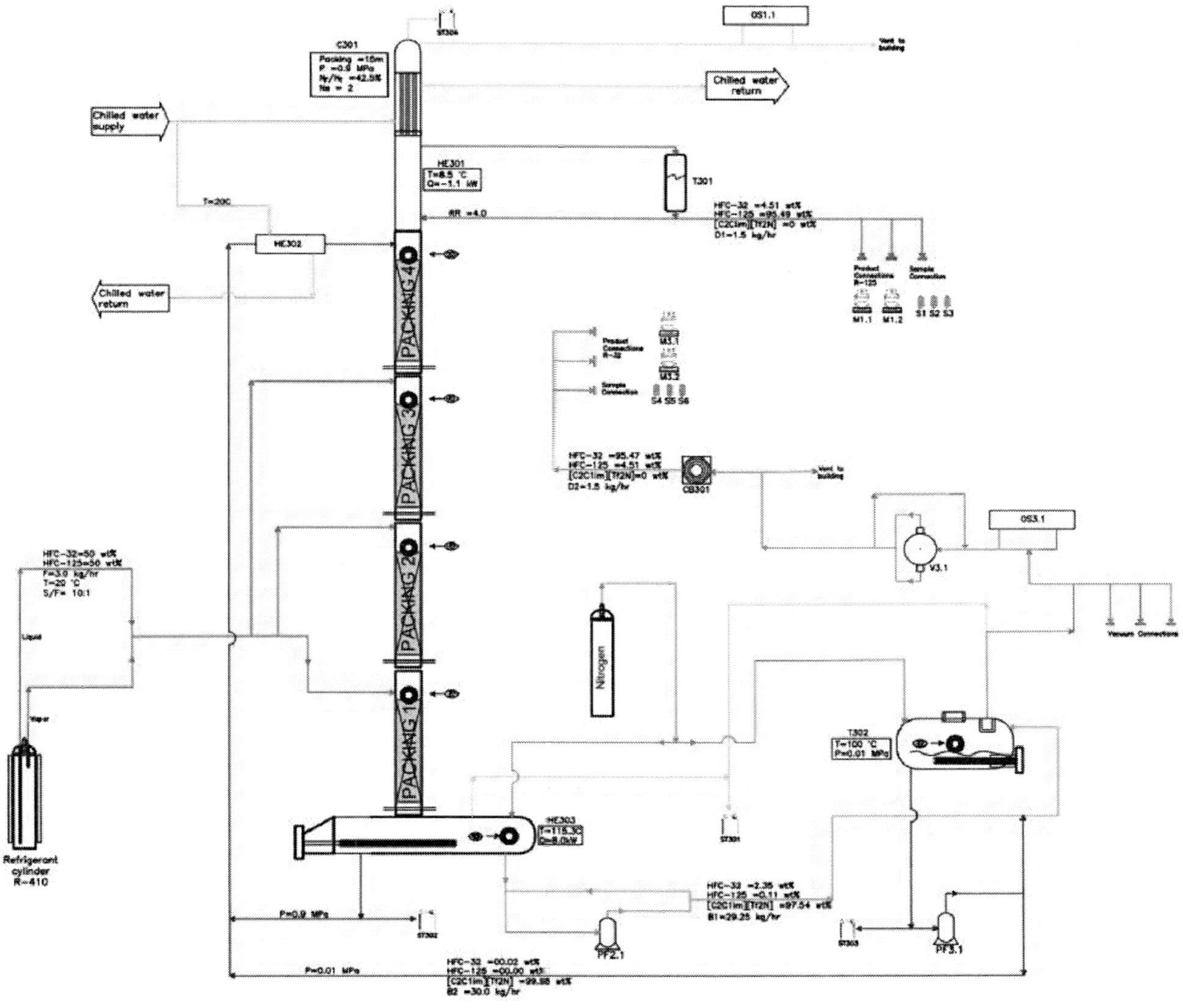

388d High Throughput Screening of Ionic Liquids for Enhanced Conductivity

Rohit Chauhan, Rohan Sartape, Meenesh R Singh

Room temperature ionic liquids (RTILs) are well-known and applied as novel absorbents, solvents, electrolytes, catalysts, and solvents in several fields of scientific research. Since it has a unique structure, RTILs exhibit several significant properties like low vapor pressure, nonflammability, a broad range of liquids, and low melting points. However, neat RTILs have a higher viscosity than organic solvents and low electrical conductivity, which limits the application of neat RTILs in many areas. To overcome these limitations, RTIL is typically mixed with organic solvents to enhance electrical conductivity and reduce viscosity. The screening of RTILs with high electrical conductivity is a significant parameter for its various applications in batteries as an electrolyte, solar and fuel cells, supercapacitors, electrochemical devices, etc. Previous studies report the mixture of RTILs and solvents, which are capable of H-bonding, and the solvents used have a lower viscosity than RTIL. The H-bonding influences the transport behaviour of the ions in the solvent, which defines the electrical conductivity. However, few current research studies emphasized the H-bonding in a mixture of RTILs for the solvent having higher viscosities (like ethylene glycol: EG). Therefore, this study used EG as a solvent to mix RTILs to enhance electrical conductivity. EG has lower vapor pressure and high boiling point than many organic solvents, making it more compatible to apply in different applications. The RTILs were screened by the high throughput method, in which the robotic hand and 96 well plate assisted electrochemical impedance spectroscopy (EIS) method was used to measure the electrical conductivity of the mixtures and neat RTILs. The obtained electrical conductivity was fitted with suitable interpolation methods. The -OH bond interaction was measured with the FTIR analysis and will be a part of this work.

388e Required Properties of Lower Critical Solution Temperature (LCST) Mixtures for Use in a Dehumidification and Cooling Cycle

Jordan Kocher, Akanksha Menon, Shannon K. Yee

Certain mixtures possess lower critical solution temperatures (LCSTs), such that they form homogeneous mixtures at temperatures below the LCST and phase separate at temperatures above the LCST. Various ionic liquids (ILs) have LCST phase behavior when mixed with water, whereby heating the mixtures above their LCSTs results in a "water-rich" and "water-scarce" phase. This enables a thermodynamic cycle wherein the water-scarce phase absorbs moisture from indoor air, providing dehumidification, and the water-rich phase desorbs moisture to the outdoor air, which both balances the mass within the cycle and cools the indoor air (through indirect evaporative cooling). However, while various IL/water mixtures possessing LCSTs have been discovered, their properties are insufficient for this cycle to be implemented in practice. Namely, the enthalpy and entropy of mixing is too small in magnitude to result in a sufficiently low chemical potential within the water-scarce phase to dehumidify building air to the low humidities required for thermal comfort. As such, the success of this new thermodynamic cycle relies heavily on the discovery of new LCST mixtures with more negative enthalpies of mixing (stronger bonds between unlike molecules) and more negative entropies of mixing (ILs that become more ordered in the presence of water). If such mixtures are discovered, the "LCST cycle" will be capable of providing dehumidification and cooling with zero global warming potential refrigerants, no water consumption, and low-grade, inexpensive heat (*e.g.*, solar heat) as the energy source.

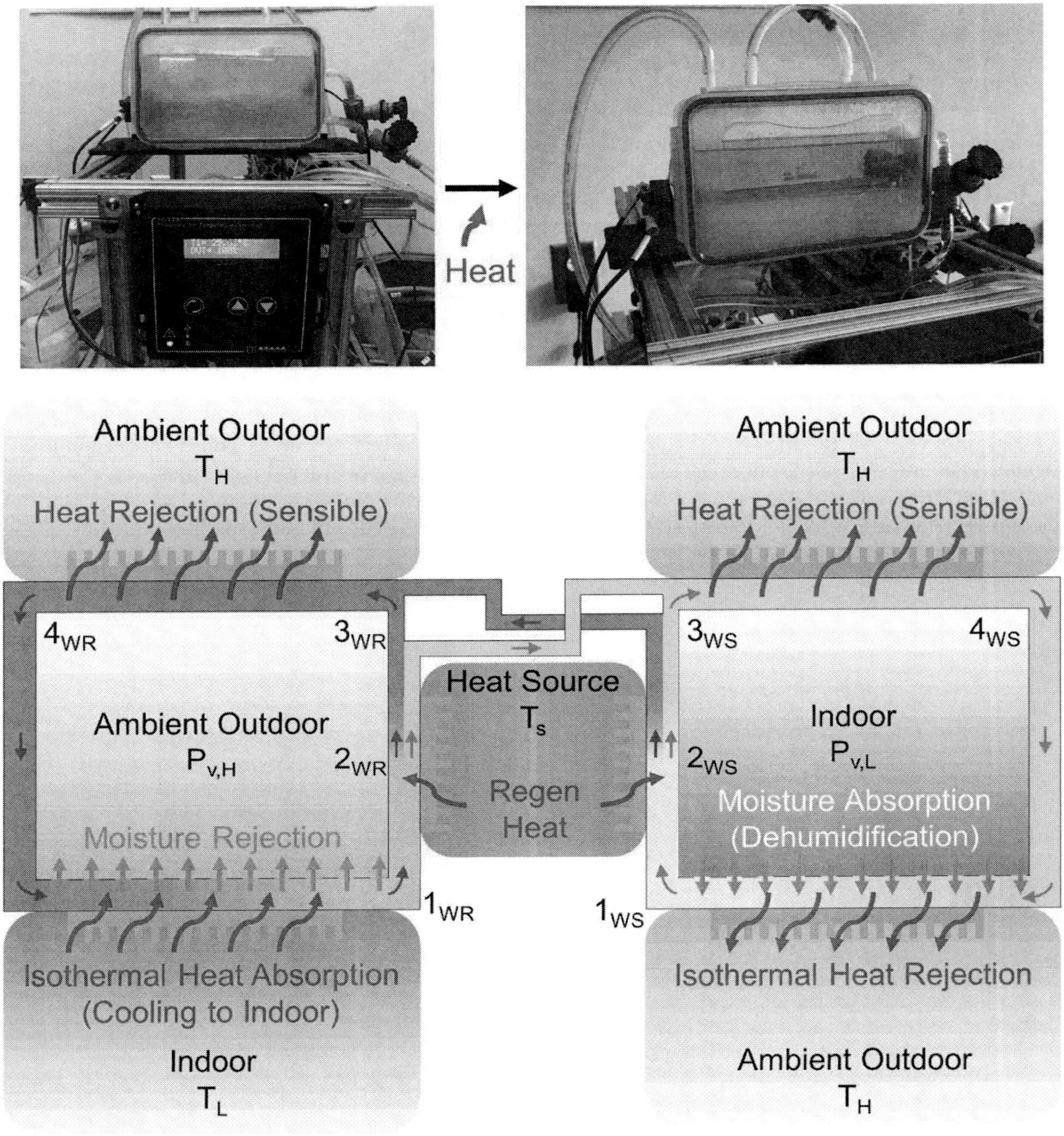

WR: Water-Rich Phase; WS: Water-Scarce Phase

388f Mechanism, Energetics and Molecular Descriptors Governing CO_2 Chemisorption in Choline Based Eutectic Solvents

Ruth Dikki, Muhammad Zeeshan, Eda Cagli, Sanchari Bhattacharia, Stephen Vicchio, Sanduni Wijesooriya, Rachel Getman, Burcu Gurkan

Eutectic solvents based on choline salts with amine functionalized anions as hydrogen bond acceptors (HBAs), and three different hydrogen bond donors (HBDs), namely ethylene glycol (EG), propylene glycol (PG), and monoethanolamine (MEA) were developed as alternative solvents for CO_2 capture at low partial pressures. Composition-dependent physical and thermal properties as well as CO_2 chemisorption mechanism and energetics were investigated. Further, simple structural descriptors of choline salts and HBDs were defined using DFT calculations and utilized in the quantitative structure–property relationship (QSPR) analysis. Accordingly, multi-linear models were developed to determine viscosity, density, and CO_2 capacity of the eutectic solvents; these models can serve as a starting point for future design of improved eutectic sorbents. In parallel, NMR analysis and DFT calculations were performed to examine CO_2 binding and reaction energetics. It was found that in eutectics with EG and PG, CO_2 binds to multiple sites including the hydroxyl oxygens on choline and ethylene glycol, and the amine on the anion. For binding to the anion, the calculated reaction enthalpies are on the order of -21.8 kJ/mol to -65.7 kJ/mol with the inclusion of the HBD. However, in the case of eutectics with MEA, it was found that the binding of CO_2 to the amine on MEA was proceeded with a hydrogen transfer from the MEA^+-CO_2^- zwitterion intermediate to the anion. This was in place of the known proton transfer from the zwitterion to another free MEA, thus achieving equimolar uptake of CO_2 per MEA in the eutectic solvent. The binding energies, which include the interaction energy between MEA and the HBA anions, ranged from -94.7 kJ/mol to -151.8 kJ/mol for CO_2 binding to MEA, with the different binding energies dependent on the proton transfer energy from MEA to the HBA anions. The lower the proton transfer energy, the more downhill CO_2 binding for MEA. As the MEA-MEA interactions are weakened by the presence of the choline salts, the extent of hydrogen bonding is lowered. This is specifically true compared to the EG and PG where CO_2 diffusivities, as examined by MD simulations, are lower than the MEA based eutectic. This presentation will discuss the roles of the HBAs and the HBDs on the CO_2 reaction mechanisms for applications in CO_2 separations.

388g Thermodynamic Modeling of Gas Solubility and Transport in an Ionic Liquid Employed As Sweep Solvent during Methanol Synthesis in a Membrane Contactor Reactor

Mohammad Bazmi, Hamid Bakhshizarinabadi, Sheng Hu, Fei Meng, Yongchun Tang, Theodore Tsotsis, Kristian Jessen

Methanol (MeOH) is an important compound finding applications today as fuel and important raw material in the chemical industry. Methanol is produced industrially by the catalytic conversion of syngas (a mixture of $CO/CO_2/H_2$) produced either through the steam reforming of natural gas or the gasification of coal or biomass. Methanol synthesis (MeS) is a highly exothermic, equilibrium-limited process, technically challenged by both low per-pass carbon conversions (typically, ~25%) and difficulties with heat management during reactor operation.

In place of the conventional MeS processes, our team proposed recently a novel reactive separation technology that makes use of a membrane contactor reactor (MCR)[1,2]. The process employs a non-volatile, high-temperature resistant ionic liquid (IL) as a sweep fluid to remove in situ the MeS products (H_2O and MeOH). As a result, significantly improved per-pass carbon conversions and MeOH selectivity are obtained.

For optimal reactor design and operation, it is essential to have effective models for estimating the solubility and transport properties of the MeS reaction components, including $CO/CO_2/H_2/H_2O$/MeOH (both individually and in the mixture) in the IL (in this case EMIM-BF_4) at the relevant high pressure and temperature MeS process conditions. In the present study, the solubilities of the MeS components in EMIM-BF_4 were measured experimentally using a pressure-volume-temperature (PVT) system. To model and analyze the experimental behavior, the Peng-Robinson and Soave-Redlich-Kwong equation of states (EOS) were applied to account for the non-ideality of the gas phase, while the UNIQUAC and UNIFAC activity coefficient models were used to calculate the fugacity of the liquid phase in equilibrium with the gas phase. The modeling results are compared with our experimental solubility data, and the applied models are shown capable to predict the experimental behavior. In addition, the sensitivity of MeS-MCR model predictions to the accuracy of the thermodynamic models is investigated. Species diffusivities were measured both in a PVT cell (constant volume diffusion experiments) and also in a high-pressure TGA system.

The measured thermodynamic and transport properties are used to optimize the performance of the membrane reactor system. These experimental data are integrated in our in-house MeS-MCR models[1] and are shown capable to predict the experimental behavior.

Keywords: Methanol synthesis, Ionic Liquid, Solubility, Equation of State, Diffusion Measurements, Membrane Reactor

References:

1. Zebarjad, S.F., Gong, J., Li, Z., Jessen, K., Tsotsis, T.T., "Simulation of Methanol Synthesis in a Membrane-Contactor Reactor," In Press J. Membrane Science, 2022.

2. Zebarjad, F., Hu, S., Li, Z., and Tsotsis, T.T., "Experimental Investigation of the Application of Ionic Liquids to Methanol Synthesis in Membrane Reactors," *Ind. Eng. Chem. Res.,* 58. 11911, 2019.

388h Treating Brine Solutions By Osmotically-Assisted Reverse Osmosis (OARO) and Thermal-Responsive Ionic Liquids

Zhongzhen Wang, Edwin Torres Cuevas, Benny D. Freeman, Joan Brennecke

Brines such as the shale gas-produced water are highly concentrated (NaCl > 1.5 M), which exhibits high osmotic pressure that cannot be further concentrated by normal reverse osmosis. Conventional solar pond evaporation is energy-efficient, but such a process is time intensive (> 1 year). Osmotically-assisted Reverse Osmosis (OARO) uses both transmembrane pressure and a draw side sweep solution as the driving force, which is suitable for the brine concentration, but multiple stages of separations and the associated energy consumption are required. In this work, instead of using NaCl as draw solutions, we developed a more sustainable OARO process by using thermal responsive ionic liquids as draw solutions. An LCST-type ionic liquid, [P4444] DMBS, with a critical temperature of around 36 °C, was used as the draw solution at different concentrations for the water recovery from NaCl solutions between 1.5 M to 2.5 M, under 65 bar transmembrane pressure. The commercially available FilmTec™ SW-30 membrane was used for OARO applications, and a bench-scale OARO system is developed. The flux is increased from 1.4 LMH to 2.1 LMH when employing 70 wt% [P4444] DMBS as draw solution and treating 1.5 M NaCl solution. The NaCl rejection is 85%, and the back-flux of [P4444] DMBS is only ~ 1.5 g m-2 h-1. The ionic liquids can be recovered by low-grade waste heat at 50 °C, with most of the NaCl partitions into the water phase. The successful development of such an IL-based OARO process provided a new pathway towards the sustainable treatment of brine solutions such as the produced water.

388i Biorefining with Low-Cost Ionic Liquids: Chemicals, Fuels and Economics

Jason Hallett

Ionic liquids (ILs) have proven to be highly tunable 'designer solvents' capable of a wide range of exciting chemistries. However, industrial application at large scale is hampered by high solvent cost. This cost is, however, a tunable feature of the solvent itself – provided the ion selection is handled with a careful eye aimed at limiting synthetic complexity. Lowering the solvent cost will increase the attractive opportunities of ILs for bulk processing of lower cost end products – including such applications as biofuels.

One of the key challenges in biorefining is the initial separation or deconstruction of lignocellulosic feedstock into separate components. ILs offer unique advantages in this area, due to their unusual thermochemical properties. However, there are serious concerns about the economic viability of their use due to the very high cost of most ionic liquids (> $50/kg).

We have overcome this by redesigning the IL based deconstruction process to use low-cost, acidic ILs for lignin dissolution rather than cellulose dissolution, yielding filterable cellulose and a dissolved lignin for precipitation or conversion to high-value chemicals. We have found that processability of the cellulose is high and lignin recoveries near quantitative.

We use a range of 'protic' ILs, the family typically used in IL industrial processes, because their simple acid-base chemistry results in a simple and cheap synthesis, with a cost (< $1/kg) similar to common organic solvents such as acetone or toluene. This presentation will discuss how ionic liquids can be 'tuned' to control cost structure of the final solvent, and what implications this will have for the chemical processes involved. The impact of the solvent on large-scale applications, such as biomass pretreatment, will be discussed, with a focus on performance and process considerations such as how the ILs maintain solvent stability under long-term processing conditions, that they can be recovered and continue to exhibit very good performance after multiple reuses. These properties highlight that the ILs have the flexibility to be useful for a variety of downstream chemical processing techniques, and for use in other applications as well.

Ultimately, the success of this approach has led to applications in other areas, including commercial activity in biorefining, textile recycling, nanomaterials production and battery recycling.

References

[1] F.J.V. Gschwend, L.M. Hennequin, A. Brandt-Talbot, F. Bedoya-Lora, G.H. Kelsall, K. Polizzi, P.S. Fennell, J.P. Hallett, Green Chem. 22 (2020) 5032-5041.

[2] H. Baaqel, I. Díaz, V. Tulus, B. Chachuat, G. Guillén-Gosálbez, J.P. Hallett, Green Chem. 22 (2020) 3132-3140.

[3] F.J.V. Gschwend, F. Malaret, S. Shinde, A. Brandt-Talbot, J.P. Hallett, Green Chem. 20 (2018) 3486-3498.

[4] A. Brandt-Talbot, F.J.V. Gschwend, P.S. Fennell, T.M. Lammens, B. Tan, J. Weale, J.P. Hallett, J.P. Green Chem. 19 (2017) 3078-3012.

[5] C.J. Clarke, W.-C. Tu, O. Levers, A. Bröhl, J.P. Hallett, Chem. Rev. 118 (2018) 747-800.

406a Enhanced Phase Change Materials for Ocean Thermal Gradient Energy Application

Jian Liu, Brianna Friedman, Hyunjun Jung, Habilou Ouro-Koura, Aljon Salalila, Daniel Z. Deng

Thermal engine based on phase change materials (PCM) has attracted attention in unmanned underwater vehicles (UUVs) because it provides economic and efficient energy storage solutions. The goal here is to develop advanced PCM with enhanced thermal properties to accelerate the heat transfer rate and increase power generation per profile for UUV. Materials with high thermal conductivity were used as additives to enhance the thermal conductivity of paraffin mixtures which was selected as the PCM candidate due to their appropriate melting temperatures (below 18 °C) and availability. The dispersion of the additives in PCM was enhanced by modification of the surface with long carbon chain molecules. The thermal conductivity of the modified PCM was measured in liquid and solid states to quantify the enhancement and correlate the relationship with the concentration of the additives in the full operating temperature range. The impact of thermal conductivity and melting temperature of the PCM on the power generation per profile for UUV were evaluated using a COMSOL model.

406b Electrochemical Reduction of CO2/CO to C2 Products: Process Design, Modeling, and Economic Analysis

Mahinder Ramdin

Direct electrochemical reduction of CO2 to C2 products such as ethylene is more efficient in alkaline media, but suffers from parasitic loss of reactants due to (bi)carbonate formation. A two-step process where the CO2 is first electrochemically reduced to CO and subsequently converted to desired C2-products has the potential to overcome the limitations posed by direct CO2 electroreduction. We investigated the technical and economic feasibility of the direct and indirect CO2 conversion routes to C2 products. For the indirect route, CO2 to CO conversion in a high temperature solid oxide electrolysis cell (SOEC) or a low temperature electrolyzer has been considered. The product distribution, conversion, selectivities, current densities, and cell potentials are different for both CO2 conversion routes, which affect the downstream processing and the economics. A detailed process design and techno-economic analysis of both CO2 conversion pathways will be presented, which includes CO2 capture, CO2 (and CO) conversion, CO2 (and CO) recycling, and product (ethylene, ethanol, acetic acid, and hydrogen) separation. Our economic analysis shows that both conversion routes are not profitable under the base case scenario, but the economics can be improved significantly by reducing the cell voltage, the capital cost of the electrolyzers, and the electricity price. For both routes, a cell voltage of 2.5 V, a capital cost of \$10000/m2, and an electricity price of <\$20/MWh will yield a positive net present value and payback times of less than 15 years. Overall, the high temperature (SOEC based) two-step conversion process has a greater potential for scale up than the direct electrochemical conversion route.

1. Ramdin, M., De Mot, B., Morrison, A.R.T., Breugelmans, T., van den Broeke, L.J.P., Trusler, J.P.M., Kortlever, R., de Jong, W., Moultos, O.A., Xiao, P., Webley, P.A., Vlugt, T.J.H. Electroreduction of CO_2/CO to C_2 Products: Process Modeling, Downstream Separation, System Integration, and Economic Analysis. Ind. Eng. Chem. Res., 2021, 60, 17862-17880.

406c CO_2 Capture Under Ambient Conditions in Functionalized Ionic Liquids

Hua Zhao

Ionic liquids (ILs) are becoming attractive solvents for CO_2 capture due to their versatile structures, but most physisorption processes in ILs required relatively high pressures (>10 bar). IL anions have strong influence on CO_2 solubility; fluorinated anions such as bis(trifluoromethylsulfonyl)imide (Tf_2N^-) and acetate (OAc^-) tend to have strong interactions with CO_2 molecules. We designed glycol ether-functionalized ILs carrying imidazolium, phosphonium or ammonium cations along with OAc^- or Tf_2N^- anions, and determined that these task-specific ILs captured up to 0.55 mol CO_2 per mole of IL (or 5.9 wt% CO_2) at room temperature and atmospheric pressure. Alcohol dehydrogenase (ADH) is one of the key enzymes used in the cascade enzymatic process to convert CO_2 to methanol, and we found that ADH was more active in aqueous solutions of Tf_2N^- based ILs than acetate analogues. This study allows us to better understand the relationship between IL structure and CO_2 solubility, and enables us to design low-cost solvent systems for large-scale CO_2 capture under ambient conditions.

406d On the Origin of Carbon Sources in the Electrochemical Upgrade of CO_2 from Carbon Capture Solutions: Case-Studies of Ag, Au, and Cu Catalysts

Avishek Banerjee, Kangze Shen, Dongfang Cheng, Philippe Sautet, Carlos Morales-Guio

Carbon capture and utilization (CCU) is a crucial technology to mitigate climate change, primarily fueled by carbon dioxide (CO_2) emission from all sectors of our economy. One approach to CCU is reactive carbon capture (RCC), which integrates 1) the capture and concentration of CO_2 in a capture media via the reversible formation of a chemical complex with a molecular capture agent, such as an amine, and 2) the direct transformation of the CO_2 complex into a fuel or chemical, circumventing a CO_2 release step. Thermal energy used to release the CO_2 into a concentrated stream for further transformation is energy-intensive and circumventing the release step can potentially reduce capital and operational costs. However, CO_2 capture media present a large degree of speciation, where carbon exists in different molecular configurations with different energy states and barriers to reaction, which adds to the complexity of RCC systems. Thus, understanding the source of carbon for transformation under RCC conditions is one of the main challenges in the field.

Utilizing a gastight rotating cylinder electrode (RCE) cell with well-defined transport properties at the gas/liquid and liquid/solid interfaces, we have carried out the electrochemical reduction of ammonium carbamate (AC), potassium bicarbonate and a CO_2-loaded MEA solutions. In all experiments, the maximum concentration of CO reached in the headspace of the cell was proportional to the partial pressure of CO_2. A combination of first principle modeling and experimental electrochemical characterization methods reveal that unbound dissolved CO_2 is the primary carbon species being consumed during the electrochemical reduction of bicarbonate and amine-based CO_2 capture solutions on silver and gold electrodes. The same studies with copper showed corrosion of the catalyst when in contact with the amine solution. Carbon in the CO_2-absorber complex occurs to directly serve as a second carbon source only at highly negative potential. This work provides a deeper understanding of the reaction mechanisms and the nature of the reacting species at the electrode interface by quantifying the transport of species in solution and by determining the competitive reaction pathways for the electroreduction of captured and dissolved CO_2, which is necessary for the rational design of catalysts and scale-up of RCC technologies.

406e A Multiscale Approach Toward Direct Air Capture of CO_2 By Biocompatible Ionic Liquids

Bohak Yoon, Gregory A. Voth

Global warming is known to be severely caused by the elevation of carbon dioxide (CO_2) emissions to the environment, a major threat to humanity. A vast majority of these emissions originate from the consumption of fossil fuels, especially for electricity generation, making it crucial to reduce the release of CO_2 into the atmosphere. Currently, the most widely used technology for post-combustion CO_2 capture is amine scrubbing, a chemical absorption CO_2 reaction process with aqueous amine solvents. However, it has several major drawbacks, such as thermal and oxidative degradation of the solvents, corrosion leading to costly equipment replacement, and high energy utilization for reclaiming the solvents. Additionally, the makeup of lost solvents can be costly, and the degradation products can be hazardous if exposed to the environment. Ionic liquids are seen as a promising alternative for CO_2 capture through direct air capture due to their low vapor pressure and high thermal stability. Specifically, amino acid ionic liquids (AAIL) have additional advantages due largely to their biocompatibility, cost-effective synthesizability, and large chemical tunability. Understanding structural flexibility in AAIL can be especially important for the chemisorption process involving chemical reactions. A few studies have attempted to reveal the reaction mechanisms for CO_2 chemisorption by AAIL using density functional theory (DFT) calculations, but the molecular mechanisms remain unclear and cannot be fully explained by static DFT calculations with only a few explicit molecules in the gas phase. A detailed understanding of dynamic features during chemisorption is urgently needed. In this work, we investigate important dynamic and kinetic aspects of the chemisorption process involving proton transport, which can be a major factor governing the overall CO_2 direct air capture process in condensed phases such as AAIL, based on *ab initio* molecular dynamics with enhanced sampling methods. Our analysis based on free-energy sampling reveals both thermodynamic and kinetic favorability for all possible reaction pathways from the zwitterion intermediate to carbamate products during the chemisorption. We further examine the intermolecular interaction and kinetic features that may play a vital role in the CO_2 reaction process. This work highlights the critical kinetic features of the direct air capture of CO_2 by ionic liquids from explicit *ab initio* free-energy sampling.

455a Development of Cellulose-Based Composite Adsorbents for CO_2 Capture

Marcellin Premila Jerome, Suresh Kumar Reddy Kuppireddy, Georgios N. Karanikolos, Nahla Alamoodi

Currently, a major environmental concern across the world is the increase in carbon dioxide (CO_2) emissions due to anthropogenic activities arising from agriculture, transportation, and industries such as cement, petrochemical, iron, and steel. On the other hand, the accumulation of waste paper in landfills is a growing concern that needs to be addressed. Several CO_2 reduction technologies are currently being investigated to offset the negative impact caused due to CO_2. Among them, the most prominent technology has been Carbon Capture, Utilization, and Storage (CCUS). At present, absorption using liquid amines are being utilized widely in industries to capture CO_2 from flue gas but it has various limitations such as high regeneration energy requirements, high corrosion, and limited CO_2 loading. As a result, alternate methods are being studied extensively.

This research aims at exploring the use of cellulose-based solid adsorbents in capturing CO_2 by reusing and recycling waste paper. Office paper comprises 85% cellulose which is extracted using an alkali and bleaching treatment in this study. Alkali treatment is done to remove hemicellulose while bleaching eliminates lignin. Cellulose is the most abundant biopolymer in the world which is inexpensive, non-toxic, and can be easily functionalized with CO_2-philic functionalities due to the presence of abundant hydroxyl groups. To further enhance the CO_2 uptake of cellulose-based materials, by creating more functional groups, higher specific surface area, and porosity; aerogels are synthesized, and the pristine cellulose is combined with graphene oxide (GO) which is a 2D material comprising hydroxyl, alkoxy, carbonyl, and carboxyl functional groups. In this study, the aerogels are prepared by dissolving cellulose in a NaOH/urea/ DI water solvent which is subsequently solvent exchanged with organic solvents and vacuum freeze-dried. The synthesized cellulose-GO aerogels were evaluated for CO_2 adsorption as a function of temperature and absolute pressure. Furthermore, the cellulose-GO aerogels were functionalized with an amino silane called 3-aminopropyltriethoxysilane (APTES) to study the effect of introducing different concentrations of APTES in adsorbing CO_2 at different temperatures and absolute pressure. To functionalize cellulose and GO, the wet gel is grafted with APTES dissolved in a hydroalcoholic mixture. By functionalizing the aerogel with APTES, it is observed that the aerogel can uptake a good amount of CO_2 at low pressures which makes the material promising for low-concentration CO_2 capture such as Direct Air Capture (DAC) applications. The study has analyzed the effect of different GO and APTES concentrations on CO_2 uptake. The optimized non-functionalized cellulose-GO aerogel displaying physisorption behavior has an uptake of 0.48 mmol/g at 1 bar and 25°C. On the other hand, the improvised APTES grafted cellulose-GO aerogel which presented chemisorption characteristics has a CO_2 uptake of 2.52 mmol/g at 1 bar and 25°C. Figure 1 represents the adsorption isotherm of cellulose, cellulose-GO, and cellulose-GO-APTES aerogels.

The cellulose-GO aerogel and the APTES functionalized cellulose-GO aerogels had also been studied for their kinetics, regeneration, and cyclability for up to 10 cycles by using both Pressure Swing Adsorption (PSA) and Temperature Pressure Swing Adsorption (TPSA), selectivity to nitrogen (N_2), water adsorption and CO2 adsorption at humid conditions. In addition, the CO_2 adsorption of the aerogels had been tested at 40°C and 60°C. As expected, the APTES functionalized cellulose-GO aerogels had higher CO_2 uptake at higher temperatures while the non-functionalized cellulose-GO aerogels had lower uptake as the temperature increased. Furthermore, the functionalized samples had much faster kinetics, higher selectivity against N_2, and showed good regeneration ability while performing TPSA. Due to the hydrophilic nature of both cellulose and GO it was observed that the non-functionalized sample had a higher water adsorption capacity and the CO_2 uptake in humid conditions at 25°C dropped to 0.4 mmol/g at 1 bar. However, due to the APTES grafting the aerogel had some hydrophobic nature due to which it adsorbed much lesser water than the non-functionalized sample. At humid conditions, the APTES functionalized sample had a lesser CO_2 adsorption capacity than the sample tested at dry conditions however, the uptake was still much higher than the cellulose-GO aerogel.

Other crucial parameters such as surface area, pore volume, pore size, and heat of adsorption have also been estimated. It is observed that the synthesized aerogels are mesoporous in nature. An important observation that was made is that the cellulose-GO aerogel had a much lower surface area than the APTES functionalized sample. This is because the cellulose-GO aerogel had a much higher average pore size and with the introduction of APTES the average pore size decreased thus, as a result of having a greater number of pores with a smaller size the surface area increased. The heat of adsorption for the aerogels was estimated using the Clausius-Clapeyron equation and the value obtained for the APTES functionalized samples was 35.95 kJ/mol while the non-functionalized sample gave a value of 14.28 kJ/mol. Both values represent how chemisorption requires higher heat of adsorption than physisorption and are within the range reported in the literature. Finally, the modeling of the CO_2 isotherms at 25°C was done using Langmuir and Freundlich models. From the curve fitting it has been observed that for the cellulose-GO aerogel both the Langmuir and Freundlich models were promising. On the contrary, for the APTES functionalized samples the Langmuir model overestimated the CO_2 capacity at low pressures and underestimated the values at higher pressures. But by using the Freundlich model the error associated with the adsorption capacity was much lesser and "n" which represents the affinity between the adsorbent and the adsorbate had a value of 5.32. Since a value of n between 2 to 8 represents high affinity and heterogenicity of the sites in the adsorbent it is concluded that the functionalized adsorbent is better modeled using the Freundlich model.

In conclusion, this works focuses on the development of a cellulose-based composite adsorbent for CO_2 capture by recycling paper and functionalizing it with APTES which is an area that has not been explored. The synthesized aerogel which is functionalized displays promising CO_2 adsorption at different conditions and when compared with similar such cellulose-based adsorbents the values are consistently making it applicable for both low CO_2 concentration sources such as DAC and industrial emissions capture.

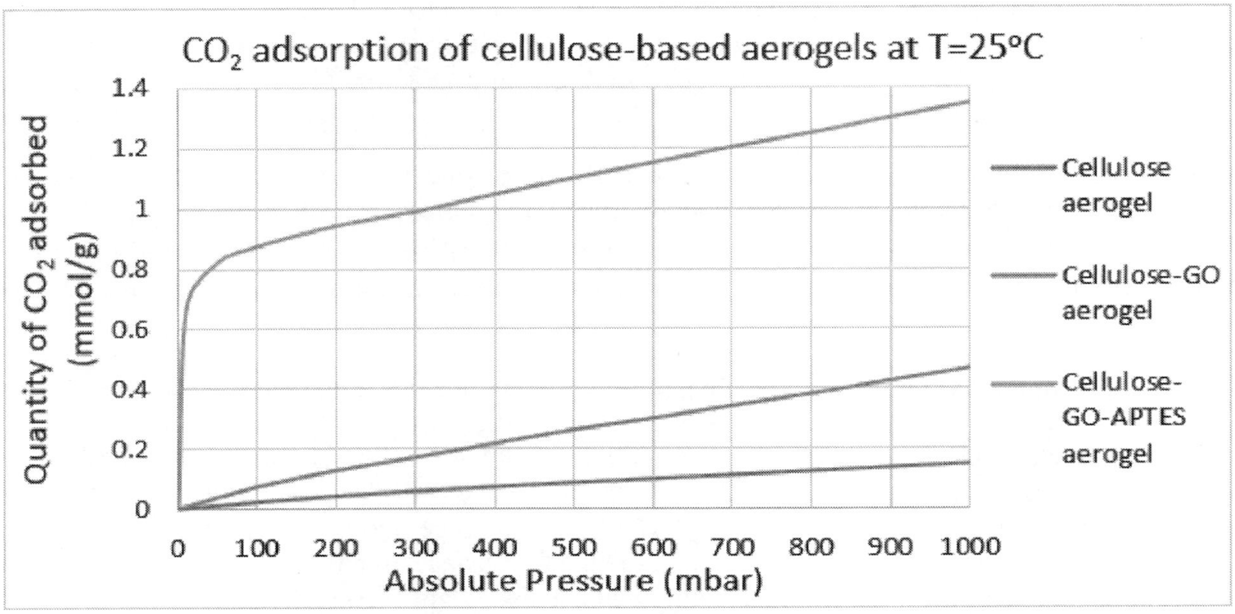

Figure 1: CO_2 Adsorption isotherm of cellulose-based aerogels at 25°C

455b Development of Double-Functionalized Halloysite Nanotubes with Enhanced Adsorption Capacity for Carbon Capture

Vijay T. John, Borui Wang, Oluwole Ajumobi, Julia A. Valla

The industrial approach to remove CO2 involves the use of solvent-based amines but the amine regeneration step has a high energy requirement due to energy wasted in solvent heating. This has led to the development of solid amine-based adsorbents where most of the energy costs are associated with regeneration of the amines. Solid amine adsorbents capture CO2 molecules through a reversible reaction between the amine functional groups and the CO2 molecules to form carbamates in dry conditions and bicarbonates in humid or wet condition. Common solid amine adsorbents can be prepared by wet impregnation of polymeric amines within the pores of a porous support (class I adsorbents) or covalent grafting of amine functional groups onto the surfaces of a porous support (class II adsorbents). We describe a method of incorporating both class I and class II adsorbents using a novel, sustainable tubular nanoclay, halloysite. Halloysite nanotubes (HNT) are naturally occurring materials with a tubular morphology where the lumen is of the order of 15-30 nm and the length is of the order of a micron. We show that the lumen can be used to encapsulate polymeric amines such as polyethyleneimine which is the class I material and the inner and outer surface of the lumen can be functionalized with aminosilanes which serve as the class II materials. The morphology of this hybrid HNT adsorbent is such that the tubular surfaces are composed of chemically attached amino groups while the HNT lumen is loaded with PEI. The results show that the capture capacity of the hybrid adsorbent is equivalent to or higher than the combined adsorption capacity of the individual class I and class II adsorbents. The capacity can be further enhanced through doubling the surface area of the nanotube by controlled acid etching. In recent work, we also show the integration of such nanotubes into highly porous materials such as MCM-41, where the HNT serves as nanostraws to allow improved access to the interior of MCM-41 and thus enhance capture capacity.

Recent lab publication:

Farinmade, A., et al. (2022). 'Tubular Clay Nano-Straws in Ordered Mesoporous Particles Create Hierarchical Porosities Leading to Improved CO2 Uptake.' Industrial & Engineering Chemistry Research 61(4): 1694-1703.

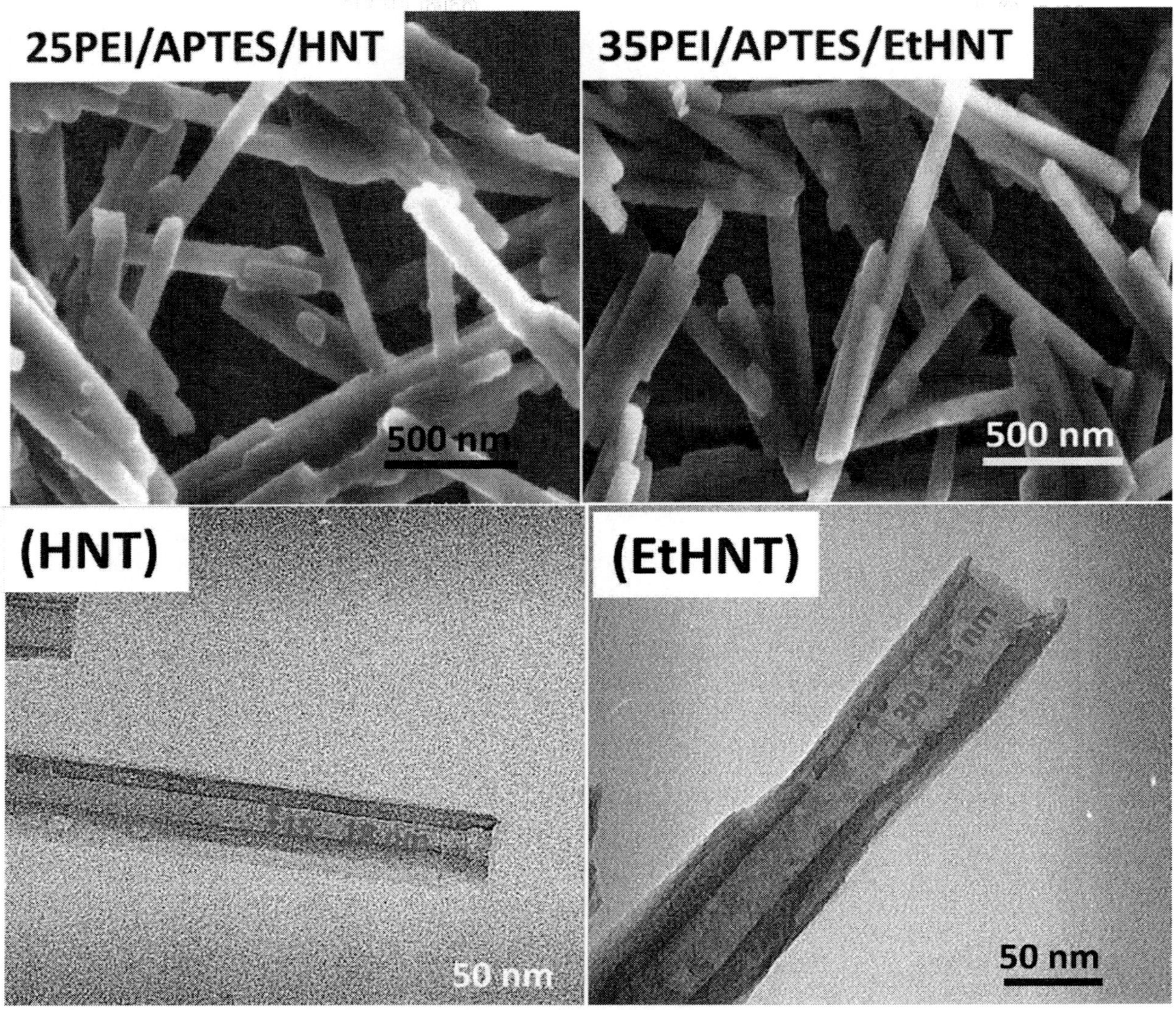
25PEI/APTES/HNT
500 nm
35PEI/APTES/EtHNT
500 nm
(HNT)
50 nm
(EtHNT)
50 nm

455c Low-Temperature CO_2 Capture from Flue Gas Using Alkali Metal Carbonate-Based Solid Sorbents in a Magnetically Stabilized Fluidized Bed Reactor

Ashin Sunny, Ishani Kudva, Pinak Mohapatra, Anuj Joshi, Liang-Shih Fan

Although renewable resources are being increasingly used for power generation, fossil fuels such as natural gas will continue to play a significant role in the foreseeable future. However, burning fossil fuels emits carbon dioxide (CO_2) into the atmosphere, a significant contributor to greenhouse gas emissions. CO_2 levels have risen rapidly since the Industrial Revolution due to growing anthropogenic activities, resulting in unwarranted climate change. This is one of the biggest problems the world is currently facing, which calls for an urgent need to develop reliable, scalable, and cost-effective CO_2 capture technologies for Natural Gas Combined Cycle (NGCC) power plants.

There are three main approaches to capturing CO_2: solvent-based, sorbent-based, and membrane-based. Currently, solvent-based technologies are considered state-of-the-art for capturing carbon emissions from NGCC power plants. However, sorbent-based technologies have the potential to be better. This method involves using a solid material to adsorb and then release the CO_2, either through temperature or pressure changes. Compared to solvent-based methods, sorbent-based technologies are less energy-intensive, can operate at a broader range of temperatures, and are easier to handle and dispose of, resulting in less environmental impact. Alkali metal carbonate-based solid sorbents are commonly used for post-combustion CO_2 capture due to their low cost and high CO_2 sorption capabilities. However, there are challenges with using this technology, such as pressure drops across the bed, and inefficient contact between CO_2 and the sorbent material. With the development of more efficient and stable sorbents along with a technology that enables good mass and heat transfer characteristics through efficient contact of gas and solids, would allow the solid-sorbent technology to compete with solvent-based methods. A possible solution is to use a magnetically stabilized fluidized bed reactor with a magnetically susceptible bed, which improves CO_2 uptake by providing efficient heat and mass transfer through bubble-less fluidization.

This study explores using potassium carbonate (K_2CO_3)-based solid sorbent for capturing carbon dioxide from simulated flue gas streams. The gas contains 4% CO_2, 10% water, and is balanced with an inert gas. The sorbent picks up CO_2 from 50°C to 90°C and can be regenerated at mild temperatures of 100°C to 150°C. The active material, K_2CO_3, is impregnated into the pores of metal oxide-based supports such as ZrO_2, Al_2O_3, and MgO, as well as other supports such as activated carbon and zeolite 13X. This helps to disperse the active material and increase the surface area available for CO_2 sorption. The impregnated sorbents are characterized using techniques like x-ray diffraction (XRD), nitrogen physisorption, scanning electron microscopy (SEM), and transmission electron microscopy (TEM). Recyclability and stability tests are conducted using a Thermogravimetric Analyzer (TGA) with consecutive cycles of adsorption and regeneration enabled through temperature swings. The sorbents are also tested at larger scales using fixed bed reactor experiments, and the CO_2 breakthrough analyses are carried out to determine their performance. Diffuse Reflectance Infrared Fourier Transform Spectroscopy (DRIFTS) is used to analyze how CO_2 and H_2O interact with the sorbent during sorption reactions, revealing insights into the reaction mechanism. After cyclic testing, the sample is further characterized to study the stability and performance of the material for long-term applications.

To enhance the heat and mass transfer characteristics of the sorption reaction, a magnetically stabilized fluidized bed (MSFB) is tested by generating a uniform magnetic field in the bed using a Helmholtz coil. The effect of magnetization on the pressure drop across the bed is studied. Different magnetically susceptible bed materials are utilized, and lower pressure drops are observed when magnetic fields are applied as compared to cases in the absence of a magnetic field. CFD simulation incorporating magnetism is studied to investigate the hydrodynamics and heat transfer inside the bed with and without a magnetic field. K_2CO_3 sorbent designed in the previous section is then impregnated onto different magnetic materials and tested for its CO_2 capture performance in a fixed bed reactor. The optimized magnetic sorbent is then tested in a hot MSFB setup under reaction conditions and the results are compared with the non-magnetic K_2CO_3 sorbent to validate the advantage of using an MSFB. The next phase in the project is to de-risk commercialization efforts by designing and fabricating a bench scale unit to study the different modes of operation: fixed bed, bubbling fluidized bed, and magnetically stabilized bed to quantify and validate the performance of MSFB.

Besides the efforts to improve gas-solid contact in an MSFB, future work on material development would focus on studying the kinetics of adsorption and regeneration steps to facilitate a better understanding of the mechanism. Promoters would be added to improve the performance of the sorbent, and the effect of mesoporous supports with high surface areas like mesoporous alumina and SBA-15 would be studied. The effect of impurities in flue gas like SO_x and NO_x on solid sorbent and hence CO_2 uptake will be investigated.

This research work would contribute scientific and technical knowledge to the carbon capture community through a deeper understanding of solid sorbent development and using novel methods like a magnetically stabilized fluidized bed for improved capture efficiency and overall CO_2 uptake. Addressing climate change by developing low-cost carbon capture technologies is the need of the hour, and this work is a promising step toward achieving this goal by capturing CO_2 from natural gas-based power plants using inexpensive, readily available materials.

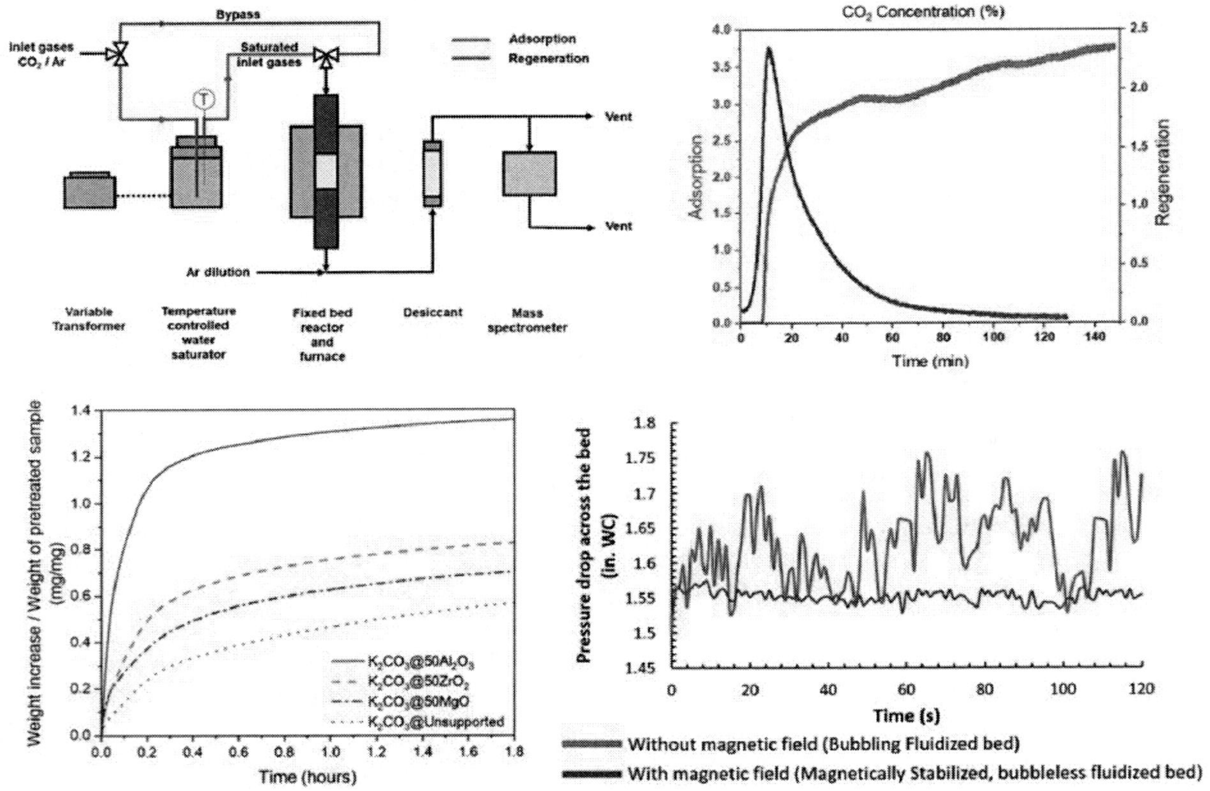

Schematic of fixed bed experiment setup (top left), adsorption and regeneration profiles for K_2CO_3@$50Al_2O_3$ in a fixed bed reactor (top right), weight increase in TGA with time for K_2CO_3@Unsupported and K_2CO_3@50(Al_2O_3, ZrO_2, MgO) during CO_2 adsorption (bottom left) and pressure drop across the bed with and without magnetization (bottom right)

455d Low Temperature Carbonation of Solid Sorbents By Reactive Milling

Seyed Morteza Taghavi Kouzehkanan, Ehsan Hassani, Farshad Feyzbar-khalkhali-Nejad, Tae-Sik Oh

CO_2 capture and storage will play a significant role in climate change mitigation. In addition to the industrialized amine-based media, CaO and $Ca(OH)_2$ have been extensively studied as promising CO_2 capture media due to their ability to form calcium carbonate. In this work the carbonation behavior of calcium-containing sorbents, CaO and $Ca(OH)_2$, was investigated under pressurized CO_2 at nominal room temperature. The carbonation reaction was mechanically driven via reactive ball milling. The carbonation rate was determined by monitoring the CO_2 pressure inside the sealed milling jar. Two different versions of CaO were fabricated as starting materials. The addition of citric acid in CaO synthesis resulted in a significant increase in sorbent surface area, bringing up the conversion of CO_2 from 18% to 41% after 3 h of reactive milling. The hydroxide formation from these two oxides closed the surface area gap. Nevertheless, we found that hydroxides had a higher initial carbonation rate and greater final CO_2 uptake than their oxide counterparts. However, the formation of byproduct water limited further carbonation of $Ca(OH)_2$. When we added a controlled amount of water to the CaO-containing milling jar, a high carbonation rate and an extensive CO_2 uptake were attained due to the in situ formation of reactive $Ca(OH)_2$ nanoparticles. We saw $CaCO_3$ X-ray diffraction peaks only when $Ca(OH)_2$ was involved in this low-temperature carbonation, indicating that the grain growth of the $CaCO_3$ is easier on the $Ca(OH)_2$ surface than on the CaO surface. Finally, CaO was in situ functionalized with 3-aminopropyl-triethoxysilane (APTES) aqueous solution during milling. This amine-based modification led to a higher CO_2 conversion. We used the Friedman isoconversional method to calculate the effective activation energy of decarbonation for the tested sorbents. The trend in the observed effective activation energies was correlated to our sorbent preparation conditions.

455e Sub-Ambient Direct Air Capture with Mixed Diluent/Aqueous Amino Acid Solvent-Based Processes

Gyoung Gug Jang, Abishek Kasturi, Joshua Thompson, Jiho Seo, Radu Custelcean, Costas Tsouris

Direct Air Capture (DAC) is an effective way to mitigate greenhouse gas emissions and can be deployed anywhere to capture carbon dioxide from the atmosphere. One of the leading DAC processes is aqueous solvent-based CO_2 absorption utilizing chemically reactive amino acid solutions that have low vapor pressure, slower evaporation rates, and reasonably high CO_2 solubility, making them suitable for large-scale deployment. However, the freezing points of aqueous amino acids that can be used for DAC are above seasonal changes in colder climates, limiting their adoption to certain geographical areas. To address this potential limitation, this study aims to develop solvent formulations for DAC that can operate within a wider temperature range, particularly at sub-ambient climate conditions, while maintaining high CO_2 solubility, kinetics, and mass transfer rates. This research has demonstrated that certain additives, such as glycols, can significantly reduce the freezing point of aqueous amino acid solvents to below -40 degrees Celsius, while maintaining CO_2 solubility. Relevant physical properties of the new solvent system such as freezing point, viscosity, density, and regeneration energy, which impact the operational implementation, CO_2 removal performance, CO_2 capacity, and reaction kinetics in cross-flow absorption systems, have been investigated. This research provides fundamental physical property data and understanding of DAC solvent deployment that are directly applicable to the scale-up of this DAC technology.

An Experimental Investigation of Microwave Desorption of CO_2 for Direct Air Capture

Rahim Boylu[1], Mustafa Erguvan[1], Shahriar Amini[1]

1. University of Alabama, Department of Mechanical Engineering, Tuscaloosa, AL, USA

Abstract

Greenhouse gas emission has been one of the most important concerns in the last three decades for humanity; and unfortunately, it continues to increase. Although several attempts have been made to reduce carbon emissions; it still requires significant efforts to minimize it. In the last couple of years, a novel method, Direct Air Capture (DAC), has received high attention to reduce carbon dioxide from air. However, DAC technology is still not mature yet; and requires a significant amount of thermal energy for the regeneration process.

While there are several methods to capture CO_2 from the flue gases, one of the most common methods, temperature swing adsorption process (TVSA), has been selected for the proposed study since it does not only require low temperature compared to the other methods but also has the advantage of having simple and less components. In the TVSA method, both adsorption and desorption processes occur in a single unit. In the conventional DAC units with TVSA method, steam is being used as a thermal energy source for the regeneration process. Due to thermal losses and long-time process heating of conventional conductive heating, the cost of desorption in the DAC system is considerably high mainly due to the desorption of CO_2.

Microwave Swing Desorption (MSD) could be an alternative way to provide thermal energy for CO_2 desorption process [1-4] compared to conventional heating process. Microwave irradiation has the advantages of volumetric heating, selective material heating, non-contact heating as well as fast heating and cooling.

In this work, experiments are performed to investigate the desorption of CO_2 under solid-state mono-mode microwave conditions using Zeolite 13X as an adsorbent material. Regeneration temperature ranging from 45 °C to 100 °C, microwave power from 5 W to 60 W at a gas flow rate of 750 ml/min, have been varied to investigate not only adsorption and desorption capacities but also energy requirements for the regeneration process. The results proved that the microwave assisted swing desorption process not only accelerated the CO_2 release but also reduced the energy consumption.

Acknowledgment

The funding from Alabama Transportation Institute (ATI) is acknowledged to carry out this work.

References

1. McGurk, S.J., et al., *Microwave swing regeneration of aqueous monoethanolamine for post-combustion CO2 capture.* Applied Energy, 2017. **192**: p. 126-133.
2. Bougie, F. and X. Fan, *Microwave regeneration of monoethanolamine aqueous solutions used for CO2 capture.* International Journal of Greenhouse Gas Control, 2018. **79**: p. 165-172.
3. Meloni, E., et al., *Intensification of TSA processes using a microwave-assisted regeneration step.* Chemical Engineering and Processing - Process Intensification, 2021. **160**.

4. van Schagen, T.N., P.J. van der Wal, and D.W.F. Brilman, *Development of a novel, through-flow microwave-based regenerator for sorbent-based direct air capture.* Chemical Engineering Journal Advances, 2022. **9**: p. 100187.

455g Production of CO_2-Negative Building Composites

Satish Nune, David J. Heldebrant, Nicholas Nelson, Keerti Kappagantula, Jaelynne A King, Yelin Ni, Jose L. Ramos, Wontae Joo, John C. Linehan, Yuan Jiang

Current state of the art carbon capture and utilization approaches often center around near-term large volume markets such as enhanced oil recovery and the production of chemicals and fuels. However, from an LCA perspective, the products associated with these markets, such as EOR and large volume fuels, are either CO_2 neutral or, in some cases, CO_2-positive. The US needs to develop new materials to create new market-driven decarbonization in the intermediate term. Despite many chemistries and markets available, CO_2 to value added functional products remain costly and many have not yet been demonstrated at a scale large enough to have a sufficient impact on CO2 emissions. We present here, an approach that combines two large-volume industrial materials (lignin/ lignite and CO_2) into a high-value building material that can aide in reducing CO_2 emissions. We detail how carboxylation of lignin and lignite creates functionalized particles that enable enhanced blending with a polymer matrix in lignin-plastic composites (LPC's). In this presentation, we present our results on the carboxylation of lignin and lignite particles and describe analytical techniques used for the quantification of carboxylic acids on lignin and lignite. We will also discuss the process parameter optimization, manufacturing, and performance evaluation of lignin polymer composites (LPC) with high filler concentrations against international building codes. We conclude with a technoeconomic and life-cycle analysis that details the cost and CO2 sequestration potential of these LPCs.

40a Systems Analyses of Green Ammonia As a Sustainable Energy Carrier

Qi Zhang

Green ammonia, which refers to ammonia produced from hydrogen obtained from water electrolysis and nitrogen obtained from air separation using renewable energy, has received tremendous attention in recent years due to its potential to decarbonize agriculture. In addition, green ammonia is being considered a sustainable energy carrier with promising applications in the energy and transportation sectors. Assessing the techno-economic viability of such technologies requires systems-level analysis and optimization at various scales. In this talk, we apply systems engineering tools to explore three ideas of how green ammonia can be used as an energy carrier: (i) Harness the strong wind on the open ocean by producing green ammonia directly offshore and subsequently shipping it to shore, which avoids the use of expensive submarine power cables for electricity transmission over long distances. (ii) Use mobile ammonia gensets for electricity generation as a non-wires alternative to improve reliability in power distribution systems. (iii) Use green ammonia as an alternative marine fuel to decarbonize the global shipping industry.

40b Sustainable Aviation Fuel from Methanol and Ethanol Intermediates

Karthikeyan Ramasamy

Conventional jet fuels derived from fossil sources not only emit substantial carbon dioxide but also contribute to global warming. Consequently, there is a pressing need for sustainable aviation fuels (SAF) that can significantly reduce emissions and promote environmental stewardship. This presentation will discuss about the processes to convert methanol (C1) and ethanol (C2) as intermediates feedstocks for SAF production, utilizing thermochemical conversion technologies. These alcohols can be sourced from a range of renewable feedstocks, including biomass, agricultural waste, and carbon capture and utilization processes. Through the employment of advanced thermochemical conversion techniques, such as dehydration and oligomerization, these alcohols can be efficiently transformed into hydrocarbon-based fuels suitable for aviation applications. The thermochemical approach offers notable advantages, including high conversion and carbon efficiency, compatibility with existing infrastructure, and the potential for large-scale implementation. This presentation will also discuss the ongoing efforts and research needs in developing SAF solutions.

40c Electrochemical Approaches to Decarbonizing Fuels and Chemicals

Haotian Wang

Electrochemical conversion of atmospheric molecules (CO_2, O_2, H_2O, N_2) into fuels and chemicals represents a green and alternative route compared to traditional manufacturing approaches. However, its practice is currently challenged at two systematic levels: the lack of active, selective, and stable electrocatalysts for efficient and reliable chemical bond transformations, and the lack of novel catalytic reactors for practical reaction rates and efficient product separation. In this talk, using CO_2 reduction to gas and liquid products and O_2 reduction to hydrogen peroxide as representative reactions, I will introduce the rational design of both catalytic materials and reactors towards practical electrochemical manufacturing of fuels and chemicals.

189a Development and Scale-up of a Novel Electrochemical Exfoliation Process to Manufacture Large Area, Few Atomic Layer, Defect-Free Graphene Flakes

William Grieco, John Hemmerling, Michael A. Miller, Phil van Wormer

In collaboration with the RAPID Manufacturing Institute® and with grant funding from the NIST Rapid Assistance (for) Coronavirus Economic Response (RACER) program, Avadain and Southwest Research Institute (SwRI) are developing and scaling a novel, environmentally-friendly electrochemical exfoliation process to convert graphite to fully reduced, large, thin and defect-free (LTDF) graphene flakes for a range of end use applications. In this presentation, the Avadain and SwRI team will describe development work to date to scale from bench to intermediate level batch production and provide insight into new patent-pending cathode assemblies that enable a step change in exfoliation productivity. The team will also offer plans for ongoing work to transition from batch to continuous exfoliation and will provide market context for the potential uses and value of LTDF graphene flakes versus other lower quality graphene offerings such as graphene oxide.

189b A Quality-By-Design Approach to Process Intensification of Green Nanomaterials

Joseph R. H. Manning, Carlos Brambila, Siddharth Patwardhan

Nanomaterials "specification" include a broad array of physicochemical features, especially if they are non- or semi-crystalline, contrasting with small molecules or bulk crystalline materials which can be provided a unique fingerprint from NMR or XRD alone. As such, their synthesis and scaleup is notoriously complex and their complex. In this study, we therefore apply the Quality by Design (QbD) paradigm during noncrystalline nanomaterial process intensification, demonstrating both its applicability to this class of materials and its importance in scaling up manufacturing.

The presentation, using bioinspired silica, will show results on how process decisions, feedstock concentration, and additive concentration affect product quality. We analysed various material properties to study the variations caused by deliberate changes to the synthesis methodology. We found that primary particle size and mass fractal dimension measured by USAXS was unable to distinguish different synthesis routes; conversely particle size distribution measured by TEM were incomparable between different synthesis methods, preventing meaningful analysis. In contrast, silica speciation and bulk textural properties (i.e. BET surface area and pore size distribution) were more sensitive to minor changes in the synthesis conditions than inter-batch variation. This helped us define the yield and bulk porosity as the critical quality attributes (CQAs), which were sensitive to synthesis conditions.

We then defined several critical process parameters (CPPs) to the synthesis at scale and assessed their impact on the CQAs. Switching to industrial feedstock from research-grade precursor led to the introduction of broad mesopores in the resultant materials. Once these modifications had been incorporated, we intensified the synthesis (increased yield per volume) by increasing the precursor concentration (and reducing additive concentrations). Interestingly, increasing the initial precursor concentration to similar levels as current industrial processes provided yields of ca. 98 %mol. Overall, we successfully applied all the changes to synthesis to make it compatible with existing industrial silica manufacturing. We increased the specific yield from ca. 1.1 g/L to ca. 38 g/L, and reduced the additive intensity from ca. 1 g/g product to 0.04 g/g, representing a significant reduction in synthesis cost and waste production. During the process intensification, it was observed that the synthesis was prone to gelation rather than coagulation under certain conditions. These results have identified a need for mapping the effects of CPPs (including scale-up) on the formation pathways and the CQAs. Such understanding can also enable process flexibility where materials of different grades can be produced using the same platform, with minor variations to the CPPs.

189c A Multi-Vortex Micromixer Based on the Synergy of Acoustics and Inertia for Nanoparticle Synthesis

Yuwen Lu, Guorui Zhu, Wei Tan

Mixing is one of the most important steps in chemical reaction, sample preparation and emulsification. However, achieving complete mixing of fluids at high throughput is still a challenge for acoustic micromixers, which are limited by the intensity of the acoustic streaming. In this study, we proposed an acoustic-inertial micromixer based on multi-vortex synergy by introducing inertia into acoustic micromixer. The device contains side-wall sharp-edge structure and contraction-expansion array structure (SCEA) in the microchannel to enhance the acoustic streaming with inertial vortices. The mixing mechanism of SCEA was explored and the mixing process showed three modes: acoustic streaming dominant mode, acoustic-inertial synergy mode and inertial vortex dominant mode. On the basis of the 'vortex seed' provided by the contraction-expansion structure, stronger chaotic advection was produced by the synergy of acoustic streaming and inertial vortices (including Dean vortex and horizontal vortex). Rapider mixing (0.20ms) and wider operating ranges (0-3000μL/min) were achieved in SCEA at lower driving voltages compared with conventional acoustic micromixers. Finally, more homogeneous and tunable chitosan nanoparticles and shellac nanoparticles were synthesized based on this device. The micromorphology, particle size distribution and drug loading properties of the products were measured and compared. This work provides a platform for control of mixing process in specific application environments with high operational flexibility, indicating potentially wider application of SCEA in multi-functional integration of lab-on-a-chip systems.

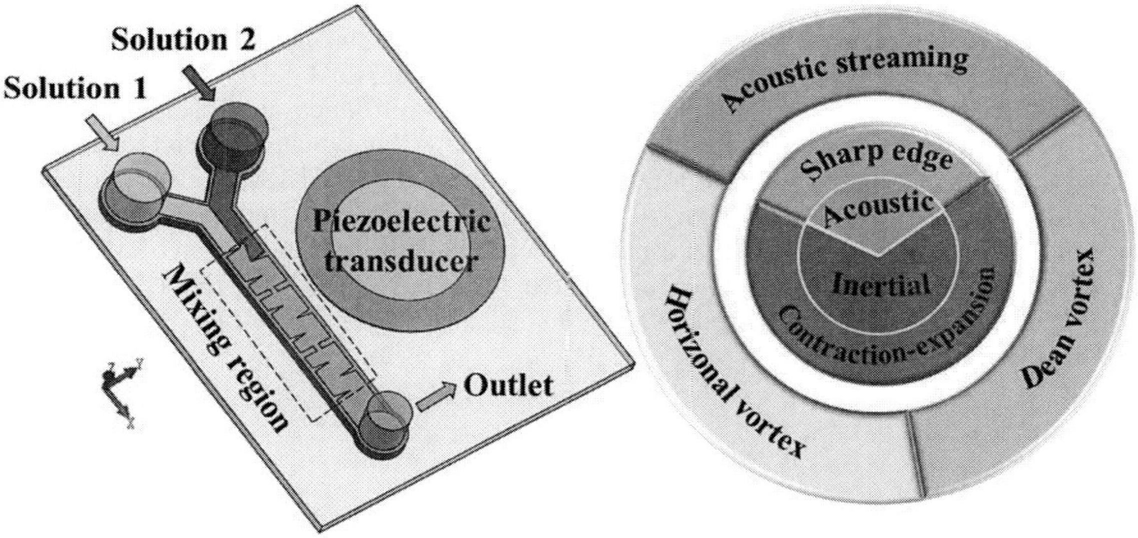

189d Development of a Modular, Continuous, End-to-End Albuterol Manufacturing Plant

Daniel Gregory, Kaitlin Kay, Justin Turnage, Brandyn Kimball, Jeffrey Noble, Frank Gupton, James K. Ferri

Conventional pharmaceutical manufacturing involves an iterative series of batch processing steps to synthesize and purify active pharmaceutical ingredients (APIs). During batch API synthesis, pharmaceutical synthons are repetitively reacted, separated, purified, redissolved, and monitored *via* chain-of-custody documentation to ensure product quality between each batch. This labor-intensive process generates large amounts of waste, necessitates large scale-up vessels which increase safety concerns, complicates logistics, and increases capital costs. These inefficiencies have reduced America's economic competitiveness in the pharmaceutical industry and has resulted in the offshoring of API manufacturing – a trend which has exacerbated supply-chain delays for lifesaving medicines. A competitive alternative methodology seeks to continuously manufacture APIs on a smaller, modular scale *via* continuous flow chemistry to facilitate high throughput conversion of API reactants into packaged pharmaceutical formulations – all within a single, streamlined, end-to-end process.

This talk explores VCU's ongoing development of a modular, continuous, end-to-end manufacturing plant for the synthesis of albuterol sulfate, a bronchodilator used for the treatment of asthma, and an API currently listed on the FDA's drug shortage list. The final automated system will be monitored using in-line process analytical technologies (PAT) to ensure product quality and facilitate the production of 120-1,200 packaged doses per hour. Key engineering design decisions necessary for successful translation of batch synthesis towards continuous manufacturing will be highlighted, with emphasis placed on model-based systems engineering (MBSE), statistical design of experiments (DOE), process intensification *via* rational synthon selection, the introduction of packed bed catalytic reactors, and operation under non-equilibrium conditions in flow. Analytical characterization *via* high performance liquid chromatography (HPLC), LC mass spectrometry (LC-MS), nuclear magnetic resonance (NMR), and spectroscopy (Raman, IR, UV-Vis) are employed to monitor albuterol synthesis in batch versus flow. The results presented herein validates the application of continuous end-to-end manufacturing as a means of exceeding batch efficiency during the production of liquid drug formulations; a strategy which can reduce capital costs and bolster America's pharmaceutical supply chain resiliency.

189e Renewable Small-Scale Synthesis of Ammonia and Its Improved Separation with Reactive Absorption

Chinomso Onuoha, Alon McCormick, Edward L Cussler, Paul J. Dauenhauer, Michael Reese, Mahdi Malmali, Tejas Nivarty, Bill Straub, Prodromos Daoutidis, Matthew Palys, Zac Pursell, Deepak Ojha, Sameer Parvathikar

Ammonia absorber columns offer an alternative separation unit to replace condensation in the Haber-Bosch synthesis loop. Metal halide salts can selectively separate ammonia from the reactor outlet gas mixture and incorporate it into their crystal lattice with remarkably high thermodynamic capacity. While the salts' working capacity can be limited and unstable when they are in their pure form, the capacity can be stabilized and increased by including a porous silica support.

Here, we discuss optimal conditions for uptake and release of ammonia. The production capacity (ammonia processed per unit absorbent and per unit production time) depends on processing parameters - including uptake and regeneration temperatures, ammonia partial pressures, release time, and sweep gas flow rate. Moreover, attention should be paid to balancing the release time with the full cycle time and bed size to ensure that the uptake breakthrough time makes efficient use of the bed. These parameters are mutually interdependent, so their optimization can be nontrivial, but rewarding. In addition, absorbent synthesis methods and thermal conductivity impact the ammonia capacity of the absorbent material as well as how quickly it can be cycled.

To address this challenge and to allow a large set of experiments, an automated dynamic absorption system was designed and fabricated. This system was used to rapidly screen uptake and release conditions, to assess material stability, and to identify optimal cycling conditions. The large amount of experimental data obtained from this automated system, and the associated mathematical models validated by this data, were found to be an essential tool to optimize the design of the absorber column. In turn, this task is essential to assess improvement in the techno-economic prospects for this new separation approach.

We also share some of our findings regarding the scaleup. These indicate potential processing challenges that need attention in design (de-risking) to ensure optimal operation in bench-scale and production-scale units.

189f CFD Modeling of a Massively Arrayed Fiber Reactor

Oluwaseyi Ayeni, Holly A. Stretz, Ahmad Vaselbehagh

The computational fluid dynamics (CFD) modeling of a static multiphase fluid reactor is described in this work. These reactors are packed with upwards of tens of thousands of fibers, and the fluid phase flows in a microfluidic environment around those fibers, massively arrayed. The geometry might be thought of simply as a microfluidic flow, annular, around the fiber. Annular microfluidic flows are not well described in the literature, nor how that might upscale to a massive array where the channels have the potential for turbulence between them. Yet these separators are commercially available, and well understood to be effective continuous systems to break or prevent emulsions while providing excellent high multiphase surface area for mass transfer to occur. Microfluidic flows are a well-researched subject, although the majority deal solely with single channels whereas the fiber reactor (FR) platform can be said to comprise thousands of channels. The computational modelling of microfluidic flows in single micro channels have been done in many studies and we reproduced a recent study in literature in order to validate our programing method. However, no CFD study has specifically attempted to model the flows in the massively arrayed channels that make up a FR. The aim of this research is to model the fluid dynamics and subsequently mass transfer performance of two-phase immiscible flows in the FR. Achieving the CFD model of a FR is challenging because the geometry comprises of channels packed with micron-sized rigid fibers. Hence, we present results of modelling flow around a single fiber using **symmetry** boundary conditions. We compare the results to a model geometry comprising four fibers using symmetry conditions. Under reported parameters, the phase distribution develops as a plugs of dispersed phase in the continuous phase similar to computational models in single channels without injection conditions. Phase structure in the four fiber model is shown and compared to the phase structure around a single fiber. Results are presented showing the effects of the channel wall wettability (contact angle), total flow rates (dispersed and continuous), flow rate ratio and interfacial tension on the equivalent droplet diameter of dispersed phase, slug length of dispersed phase, and specific area of the dispersed phase.

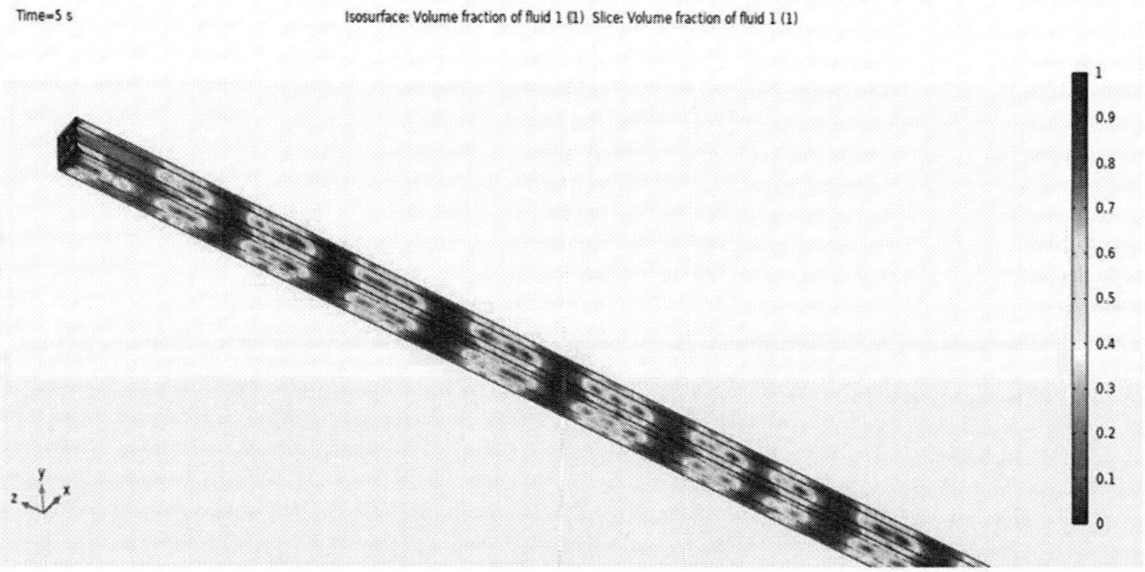

Figure 1: Phase structure around four fibers

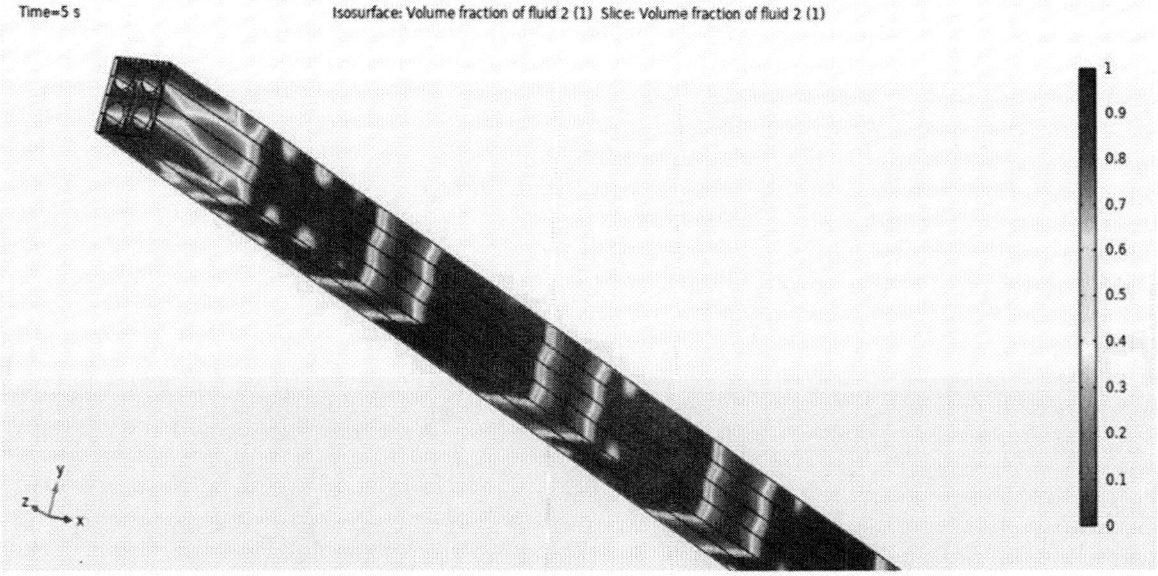

Figure 2: Phase structure around a single fiber tessellated on edges to create an array of four fibers

AUTHOR INDEX

AIChE
120 Wall Street, FL 23
New York, NY 10005-4020

ISBN 978-1-7138-9313-4

9 781713 893134